D0578750

CHEMICAL VILLAINS

a biology of pollution

Cover photo: Osprey on nest in Crane Prairie Osprey Refuge, Oregon. The osprey, *Pandion haliaetus,* is one of the species that has been most affected by the widespread use of chlorinated hydrocarbon pesticides. (Courtesy Jim Anderson, Bend, Oregon.)

JAMES W. BERRY
DAVID W. OSGOOD
PHILIP A. St. JOHN
Department of Zoology, Butler University,
Indianapolis, Indiana

CHEMICAL VILLAINS
a biology of pollution

With 89 illustrations

Saint Louis
THE C. V. MOSBY COMPANY 1974

94594

Copyright © 1974 by
The C. V. Mosby Company

All rights reserved. No part of this book may be reproduced in any manner without written permission of the publisher.

Printed in the United States of America

Distributed in Great Britain by Henry Kimpton, London

Library of Congress Cataloging in Publication Data

Berry, James William, 1935-
Chemical villains.

1. Pollution–Toxicology. 2. Environmental health. 3. Chemicals–Physiological effect. I. Osgood, David W., joint author. II. St. John, Philip A., joint author. III. Title. [DNLM: 1. Air pollution. 2. Environmental health. 3. Water pollution, Chemical. WA30 B535c 1974]
RA566.B47 615.9 73-11099
ISBN 0-8016-0663-2

E/M/M 9 8 7 6 5 4 3 2 1

628.53
B534a

PREFACE

Environment and ecology are the "in" studies of the 70's as DNA and molecular biology were during the 60's. We have chosen to omit "Environment" and "Ecology" from our title because of their recent overuse—not because this book is on a different subject. It is a book about the environment and specifically about the substances that have come to be called pollutants. Words such as sulfur dioxide, phosphate, mercury, and PCB have become common in the popular press in recent years, and it is these and many other substances that the book is about. Some of them occur naturally, whereas others are man-made, but the distribution of all of them is influenced by man.

Pollutants, or environmental contaminants, and what they do in the environment and in the plant and animal are the subjects of our book. A vast literature has developed in this field, and current research is adding new material at a great rate. Many scientists are working to determine the effects of various pollutants on both living and nonliving systems. Their results are being reported in literally thousands of technical journals, but most of their findings are inaccessible to the great majority of the population. Our goal in writing this book is to bring together some of this information in a form that can be understood and appreciated by the interested student and layman.

By their very nature environmental contaminants do not lend themselves to subdivision into neat chapters. Treating them as air pollutants, water pollutants, solid wastes, and so on makes it difficult to discuss effects of particular compounds because in many cases a compound may be equally important as a contaminant of both air and water. We have chosen to use the type of chemical compound, whether element, inorganic, or complex organic, as a means of dividing the subject matter into chapters of readable length. It is a scheme designed for convenience and has no necessary relationship to the way that these materials occur in the environment.

We hope that the book will become a source book for college students, laymen, legislators, and anyone else who has an interest in what particular environmental contaminants do in natural systems. Currently there is no single source to consult for answers to questions such as: What does sulfur dioxide do to people? Why did environmentalists seek a total ban on DDT? What are the problems in using high-phosphate detergents? In this book the reader will find answers to these questions and many more like them. We have tried to provide the information in terms that the educated layman can understand. The book can be read with profit by those with no special training in chemistry or biology, but for those with advanced training in these fields we have included more technical treatments of certain chemical and biological processes.

In many cases we have had to be content with a discussion of the symptoms produced by a particular substance because the underlying mechanisms are not yet known. In this regard the book can serve as a source of inspiration to young scientists looking for problems to solve. It also means that many sections of the book will quickly become outdated as more becomes known about how pollutants work in living systems.

Our aim has been to write a book that will be useful as a supplement to introductory courses in biology, ecology, environmental problems, and other similar ones. It is also designed for general audiences outside the classrooms; we would hope therefore that many concerned citizens will find it useful in understanding some of the problems encountered in our contemporary world.

We would like to express our appreciation to the many people who contributed to the preparation of this book. The National Technical Information Service of the Environmental Protection Agency was especially helpful in providing technical literature searches on many of the topics covered here. The Environmental Protection Agency also provided us with photographs. We would also like to thank Sherie Zahn and Chris Stewart, who provided us with most of the line drawings. Our special thanks go to Betsy Berry and Judy Osgood for proofreading and providing suggestions for the manuscript. Judy Osgood was also our faithful typist.

JAMES W. BERRY
DAVID W. OSGOOD
PHILIP A. St. JOHN

CONTENTS

PART ONE

PART TWO

APPENDIX

PART ONE

The first part of this book discusses some basic concepts involving ecosystems and biogeochemical cycles (Chapter 1), some aspects of cell physiology (Chapter 2), and some functions of tissues and organs of animals (Chapter 3). A solid appreciation by the reader of at least these three broad areas is necessary to an understanding of the material in the second part, which details many environmental contaminants and their actions on cells and tissues of organisms. The reader who has had a good grounding in biology may find this first section unnecessary reading. Those with less preparation should begin with the first chapter.

Chapter 1 defines and describes the concept of the ecosystem as an integrated and delicately balanced assemblage of plants and animals together with their nonliving environment. First, the living organism is seen as being wholly dependent on its physical, chemical, and biological surroundings. Life cannot exist in a vacuum ("no man is an island"). Second, a constant flow of energy into an ecosystem is the "force" that keeps the living engine running. When energy flow stops, life stops. Finally, the materials of life–its chemicals–are constantly cycling from organism to environment and back again. Populations of organisms increase in numbers when nutrients are added to their ecosystems; the loss of such nutrients results in reductions in population size. Losses of essential raw materials (to sediments in deep lakes and oceans) can be permanent–at least from the perspective of present populations. From these concepts comes a sobering conclusion: Man lives in a finite world. He *can* run out of raw materials for life. He *can* run out of space to deposit life's wastes.

Chapter 2 discusses the following two important concepts relating to the activities of cells: (1) the transforming of energy locked in the form of complex chemicals and (2) the method used by a cell to construct its enzymes, catalysts, and other proteins.

Energy transformation begins with the green plant cell, which through a process called photosynthesis is able to utilize the energy in sunlight to construct organic compounds from the much simpler carbon dioxide and water. Animal (and plant) cells then extract the energy from these organic compounds (by degrading them back to carbon dioxide and water). They use the energy released for all the activities that are necessary to support life. Enzyme synthesis is one of the most important of these energy-requiring activities of cells. Most chemical reactions in a cell would occur too slowly to be of value without these catalysts. This process, because of its importance, is treated in some detail. Energy transformation is nearly synonymous with life itself, and many pollutants in our environment affect energy transformations by the cell. Proteins form much of the structure, the fabric in essence, of the plant and animal body. Some pollutants also affect the chemical fabric and the reactions in the cell that make the proteins.

In Chapter 3 the animal body is examined at the tissue and organ level. Digestion of foods, breathing, transport, and communication systems (blood and nerves) are possessed by all complex animals, and each must function correctly for normal health. In the second part it will be observed that environmental contaminants—"chemical villains"—may strike at the following levels: at the ecosystem, at the plant and animal cell, and at animal (and plant) tissues and organs. Again, no organism lives in isolation. What contaminates our environment will contaminate us.

1 ECOSYSTEMS, ENERGY FLOW, AND BIOGEOCHEMICAL CYCLES

Some years ago the World Health Organization launched a mosquito control program in Borneo and sprayed large quantities of DDT, which had proved to be very effective in controlling the mosquito. But, shortly thereafter, the (thatched) roofs of the natives' houses began to fall because they were being eaten by caterpillars, which, because of their particular habits, had not absorbed very much of the DDT themselves. A certain predatory wasp, however, which had been keeping the caterpillars under control, had been killed off in large numbers by the DDT. But the story doesn't end there, because they brought the spraying indoors to control houseflies. Up to that time, the control of houseflies was largely the job of a little lizard, the gecko, that inhabits houses. Well, the geckos continued their job of eating flies, now heavily dosed with DDT, and the geckos began to die. Then the geckos were eaten by house cats. The poor house cats at the end of this food chain had concentrated this material, and they began to die. And they died in such numbers that rats began to invade the houses and consume the food. But, more important, the rats were potential plague carriers. This situation became so alarming that they finally resorted to parachuting fresh cats into Borneo to try to restore the balance of populations that the people, trigger-happy with the spray guns, had destroyed.*

This true story demonstrates how the environment is a complex of both living and nonliving factors that interact with each other. The environment has no walls or barriers between any of these factors so that it is difficult to isolate a change in one part of the environment from the other parts. This concept is called the *holocoenotic environment* (*holo* = the whole; *coenotic* = without walls) to indicate the far-reaching implications that may result from making a single change in the environment. Fig. 1-1 shows how making a change in any component of the environment will have effects not only there but at many other points as well. The interaction of these

*From Ecology: the great new chain of being, by Gordon Harrison, Natural History Magazine, Dec., 1968. Copyright The American Museum of Natural History, 1968.

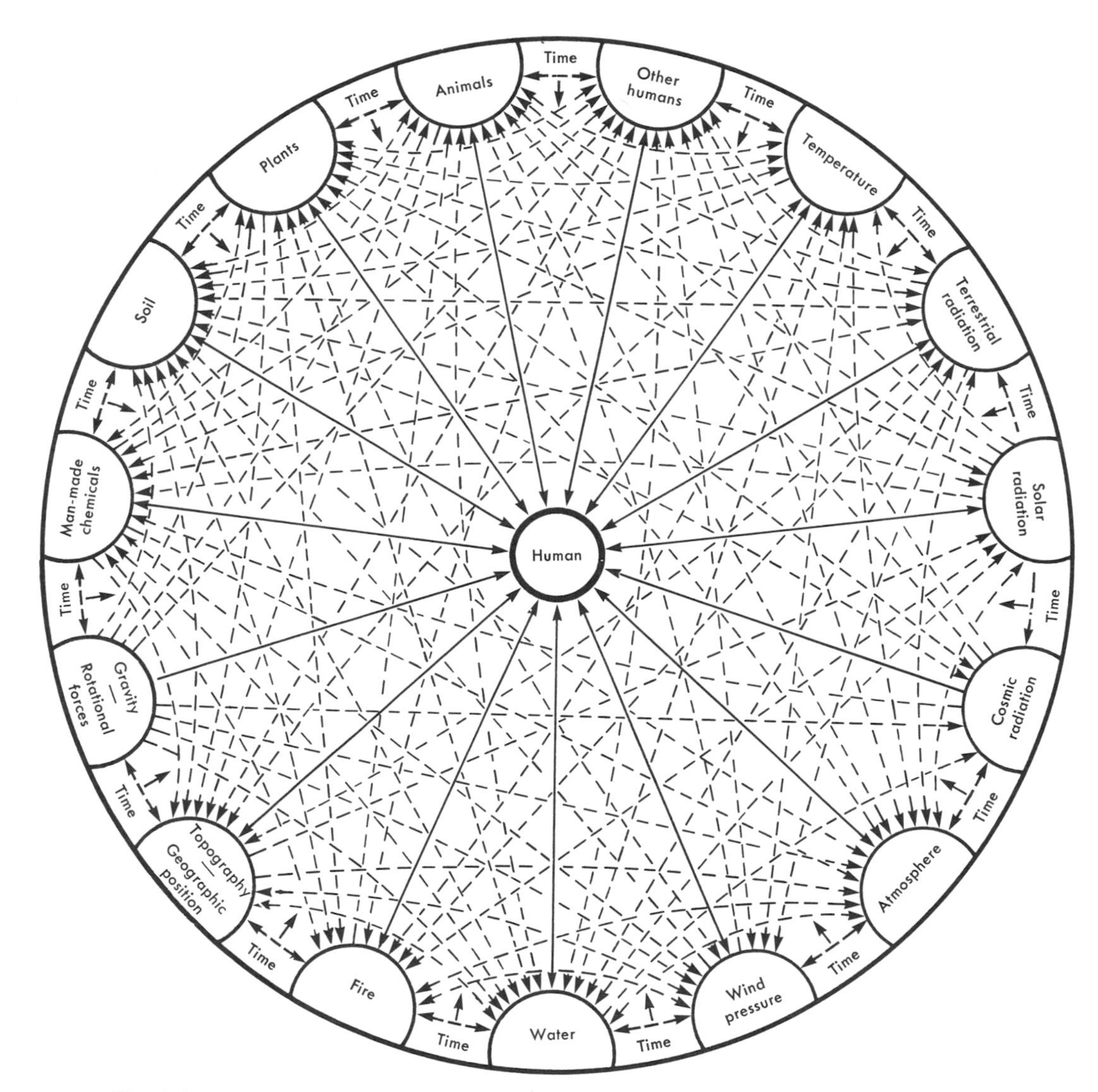

Fig. 1-1. Holocoenotic environmental complex. This diagram represents the complex and indirect interactions between an organism and the various components of its environment. The solid lines show the interaction between the environment and man, and the dashed lines show the interaction between various parts of the environment. The modifying influence of time is indicated by the inward-pointing arrows just inside the border. This diagram shows why it is impossible to make "just one change." (Modified from Billings, W. D.: Quarterly Review of Biology **27**(3):251-265, 1952.)

living and nonliving components of the environment creates an ecological system called an *ecosystem.* The study about conditions in Borneo shows how one ecosystem functions, but other ecosystems can be very large or very small. The whole world is really just one ecosystem; but on another scale a farm pond is also an ecosystem.

Although the ecosystem does function as a single entity, to understand its structure and operation it must be subdivided into components that can be experimented with and analyzed. Perhaps it is easiest to divide the ecosystem into physical components and biological components. The physical components include such factors as energy, water, minerals, gases, gravity, and countless more. Biological components include all the living organisms and the functions they perform in the ecosystem. For example, a rabbit is more than just an animal in a field. It feeds on the green plants there and provides food for foxes and other predators. All the living organisms depend on and are influenced by the habitat, temperature, moisture, sunlight, soil, minerals, and topography as well as by other organisms present.

ENERGY IN THE ECOSYSTEM

Physical principles. Energy is defined as the capacity to do work, and almost all the energy in an ecosystem originates as radiation from the sun—solar radiation. Small amounts of energy are derived from radioactive rocks, volcanoes, and similar sources, but they contribute little to the energy flow through the ecosystem. The solar radiation received at the earth's surface consists mostly of infrared (thermal), visible (light), and ultraviolet radiation (Fig. 1-2).

Radiation with wavelengths between 400 and 760 nanometers (nm) is called *light.* Other high-energy radiation is received at the upper atmosphere, but it is absorbed or reflected by the atmosphere and little of it reaches the surface of the earth (Fig. 1-3). Most of the radiation that does reach the surface is either reflected, used in evaporation, or used in raising the temperature of air, soil, and water. Only about 1% of the energy in solar radiation is actually absorbed by the plant in the photosynthetic process. (Photosynthesis is the process whereby plants containing chlorophyll transform radiant energy into chemical energy, which is stored in molecules that are used for food.) Since photosynthesis is the only way sunlight can be trapped for use in the ecosystem, all organisms (with the exception of a few chemosynthetic microorganisms) depend on the energy captured by photosynthesis.

Before discussing the role of energy in the ecosystem, it is necessary to understand some basic physical principles that apply to ecology. The *First Law of Thermodynamics* states that although energy can be neither created nor destroyed, it can be converted from one form to another. That is, the total amount of energy in a system may remain constant, but it may take on a variety of forms: energy of motion, gravitational force, elastic energy, chemical energy, heat, nuclear energy, or other forms. Energy from the sun can be converted to stored plant food, animal fat, work, or heat. Physical processes change only the *distribution* of energy.

The *Second Law of Thermodynamics* deals with the direction of energy change and can be stated in the following ways:

1. There is a decrease in the amount of useful energy whenever an energy conversion occurs. Some energy is downgraded to heat and lost.
2. Because some energy is always changed into *unusable* heat energy, no conversion of energy is ever 100% efficient.

These laws mean that when solar energy strikes the earth, it tends to be downgraded into heat energy. Only a small portion of the available energy is converted by photosynthesis into plant material. An animal eats the plant (chemical energy source), converting a large part of it into heat and storing a small amount in new animal protoplasm (chemical energy).

It should be reemphasized that the energy level of heat is so low that no further conversions are ever possible with heat energy (except when transferred

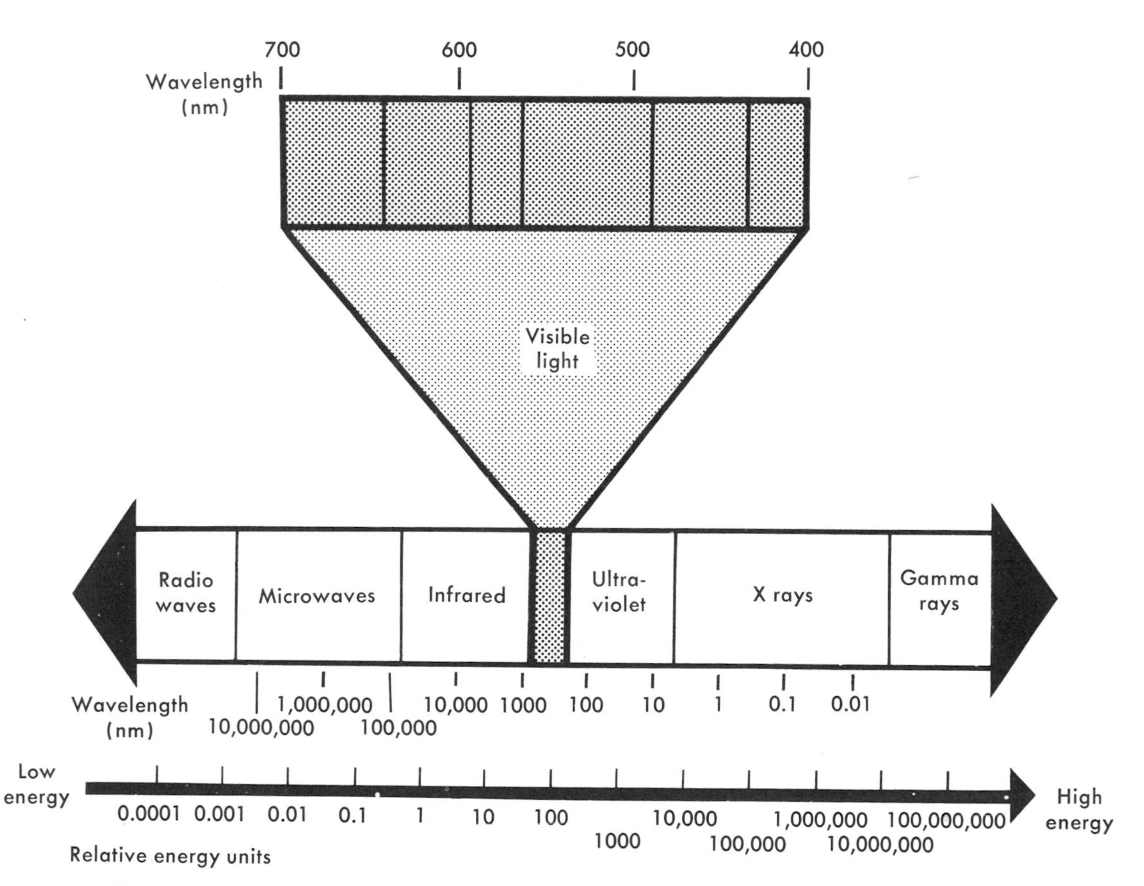

Fig. 1-2. Electromagnetic spectrum. The only difference between one kind of radiation and another is the wavelength. From gamma rays through light waves the waves range upward in length to the longest radio waves. Wavelengths longer than about 1000 nanometers (nm) are not visible but can be felt as heat. The difference in wavelength is associated with a significant difference in the amount of energy conveyed by radiation at each wavelength. The energy content is inversely proportional to the wavelength.

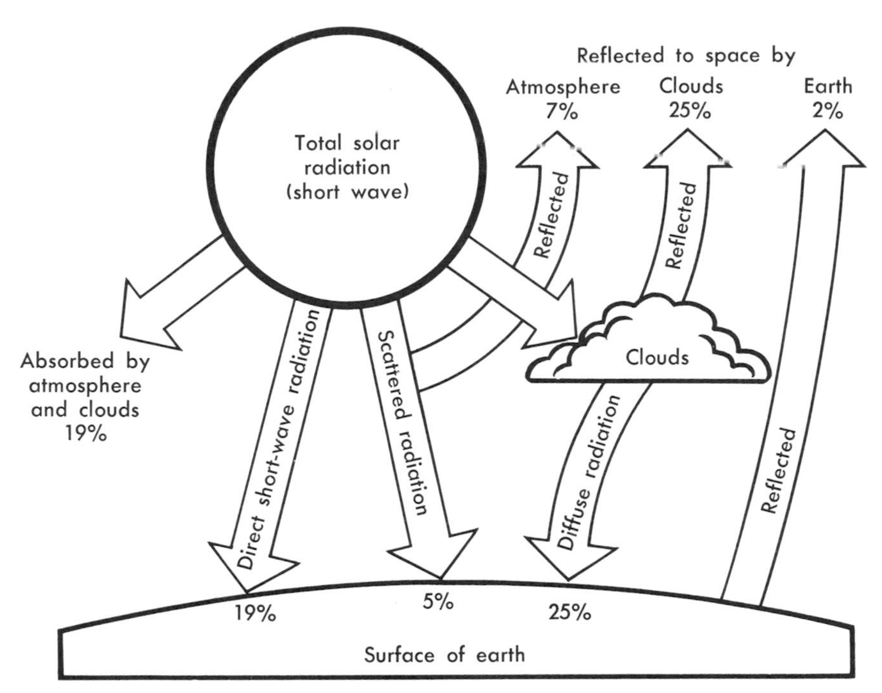

Fig. 1-3. Fate of the solar radiation reaching earth's atmosphere. Notice that a large part of the radiation never reaches the surface at all. Only that portion reaching the surface is available for photosynthesis.

from a high-temperature region to a low-temperature region). What does this mean in an ecosystem? The energy supplied to the ecosystem is constantly running downhill, away from useful energy toward useless energy. As an example, the highest energy source, solar radiation, is used by plants to produce a chemical energy source (e.g., fat, sugar), composed of carbon, hydrogen, and oxygen. These products can then be used as a food source by another living organism. After the food has been metabolized by the organism, what do we have left? The carbon, hydrogen, and oxygen that composed the food source still exist but are in the form of carbon dioxide and water, and the chemical bond energy of the sugar or fat has been released. At this stage carbon dioxide and water have reached such a low energy level that they are not useful as energy sources. Remember the Second Law of Thermodynamics! Some of the energy from metabolism of sugars or fats was used to do work or keep the organism alive, but a large percentage of it was lost as heat. (Why does a room full of people become overheated?)

Flow of energy. It has been established that the energy in solar radiation enters the living part of the ecosystem through photosynthesis. As the *producers* (green plants) grow and accumulate food energy, they are eaten by animals called *herbivores* (plant-eaters). Herbivores are then eaten by *carnivores* (meat-eaters). Eventually the carnivores will be eaten by another carnivore or die of other causes. Then *decomposers* (bacteria and fungi) will use the carnivore's body as their own energy source, leaving only carbon dioxide, water, and perhaps other waste products. This simple sequence of organisms is called a *food chain* (Fig.

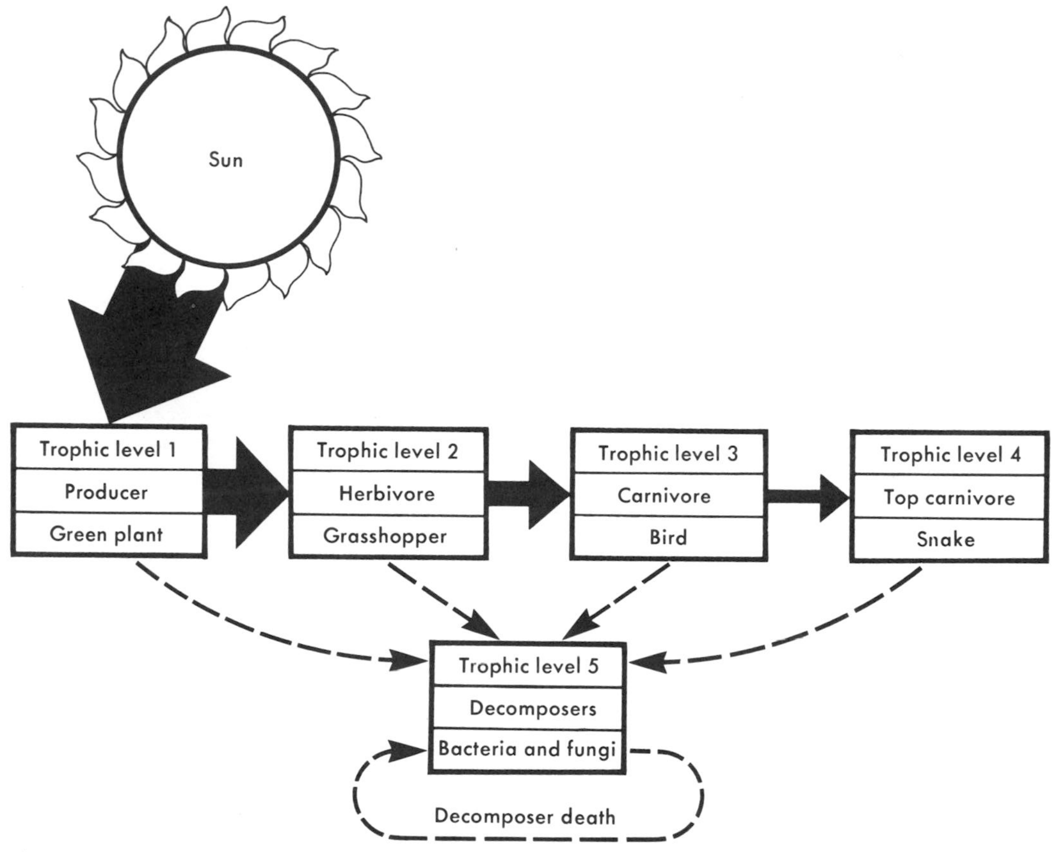

Fig. 1-4. A simple food chain. This shows the flow of energy from the sun through the plants that tie up the sun's energy and on through the various types of animals in the food chain. Any energy in dead matter is used by the decomposers. A very small amount of energy recycles at the decomposer level. As bacteria and fungi die, their tissues are acted on by other bacteria and fungi. Otherwise, the energy passes only from one level to the next higher level.

1-4), and each successive level in this chain is a *trophic level* (*trophic* = feeding). The trophic levels are numbered in sequence, with all the green plants in the ecosystem making up the first trophic level, all the herbivores making up the second trophic level, and so on up. For simplicity it is assumed that each kind of organism occupies only a single trophic level, thereby forming an unbranched food chain. In natural situations this is almost never the case, and man is no exception. As a herbivore, man's diet includes fruit and vegetables; as a carnivore, man eats beef, pork, and fowl. As a top carnivore, man eats carnivorous fish (e.g., bass, trout, salmon, tuna). Likewise, herbivores and carnivores rarely utilize only one source of food, as implied by the description of a food chain. More often there is a *food web,* with herbivores feeding on many kinds of plants, carnivores feeding on many kinds of herbivores, and

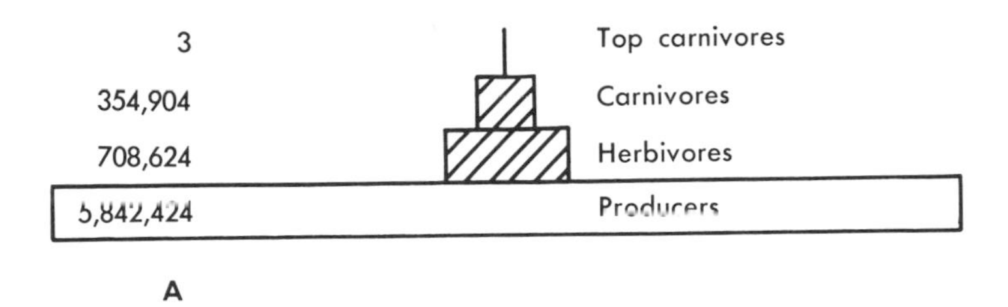

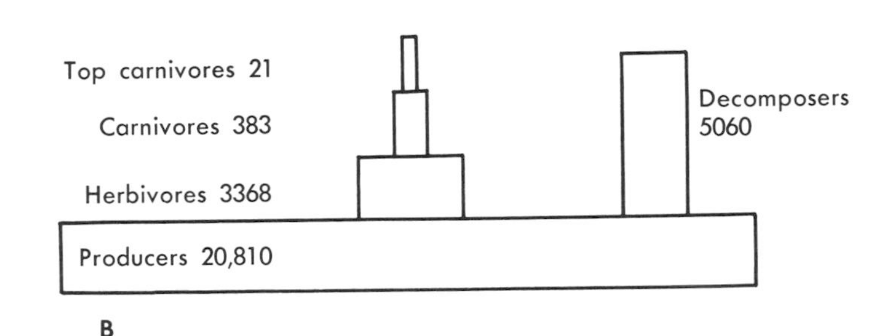

Fig. 1-5. Ecological pyramids. **A,** Pyramid of numbers for an acre of Kentucky bluegrass. This represents almost 6 million plants, over 700,000 herbivorous invertebrates, about 350,000 carnivorous invertebrates such as spiders, ants, and predatory beetles, and about three carnivorous vertebrates such as birds or moles. **B,** Pyramid of energy for Silver Springs, Florida. The numbers indicate the kilocalories per square meter per year that are actually fixed as living protoplasm at each trophic level. (**A,** Data from Odum, E. P.: Fundamentals of ecology, ed. 2, Philadelphia, 1959, W. B. Saunders Co.; **B,** data from Odum, H. T.: Ecological Monographs **27**:55-112, 1957.)

several top carnivores feeding on several kinds of carnivores. Energy passes from organism to organism, with each organism losing a certain amount of the available energy as heat.

Since the initial amount of energy in a given area is limited by the intensity and duration of sunlight, there is a point somewhere along the line where all the initial energy tied up by the plants will have been converted to heat energy. It will be recalled that, according to the Second Law of Thermodynamics, some energy is lost each time a conversion is made. This means that any food chain is limited in length because the energy supply becomes depleted. Perhaps a simple analogy will help to clarify this rather complex concept. If there is a line of buckets and water is poured from the first bucket into the second bucket, and so on down the line, water can still be poured into the last bucket, no matter how long the line is. Suppose, instead, that each bucket had holes in it which let some water leak out. The buckets still hold water, but they each lose a certain amount so that each successive bucket contains a little less water than the previous one. A point is soon reached where no water is left. Even if one starts with a larger bucket, it will have more holes in it so that one is really not much better off. In this analogy each bucket represents a living organism at a given trophic level, and the leaking water represents the energy lost by the organism through energy conversions, which is known as metabolism.

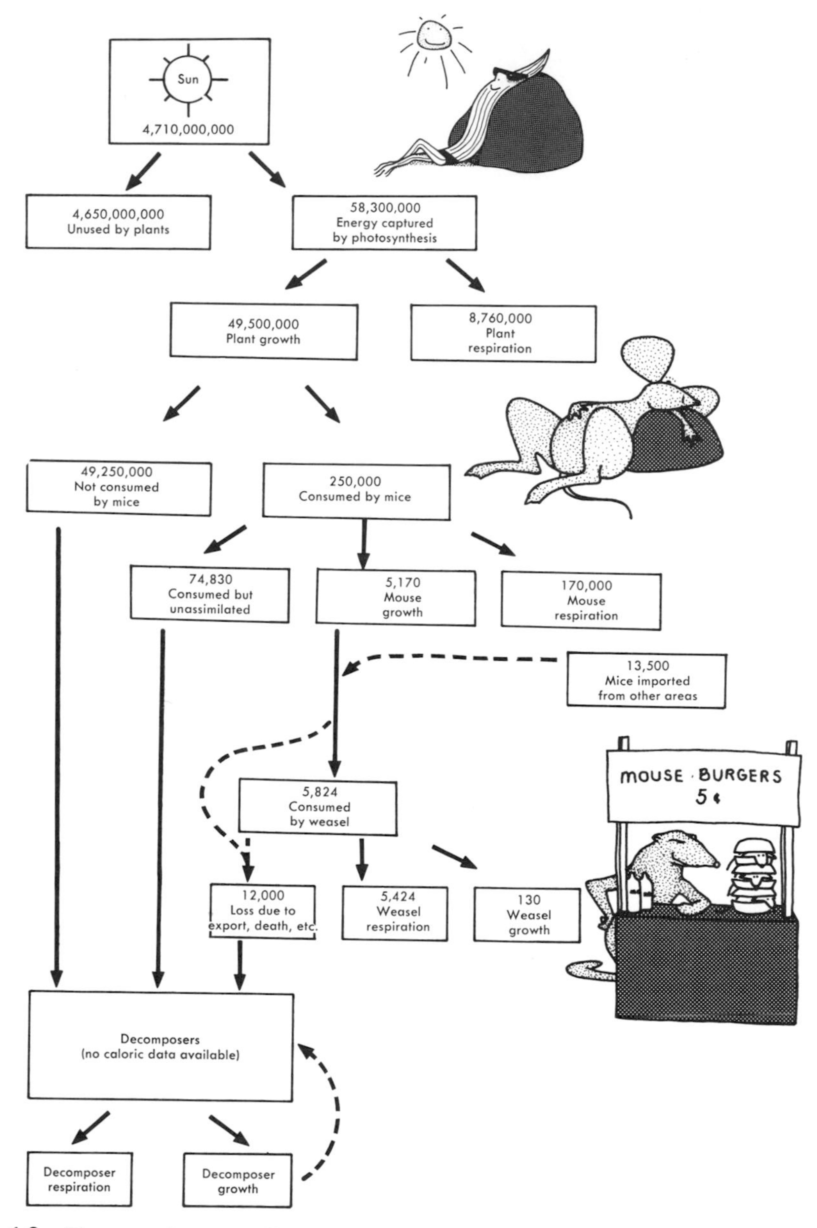

Fig. 1-6. Diagram of energy flow for a plant–meadow mouse–weasel food chain in an abandoned field ecosystem in Michigan. The principal plant was Canada bluegrass, the principal herbivore was the meadow mouse, and the principal carnivore was the weasel. All figures are in kilocalories per hectare. (A hectare is equal to approximately 2½ acres.) Since the weasel consumed more kilocalories than the mice produced, it is probable that the weasel had a food source other than mice. (Data from Golley, F. B.: Ecological Monographs **30:**187-206, 1960.)

The number of organisms at each trophic level in a food chain is determined by the efficiency with which energy is passed from one level to another. Under favorable conditions only about 1% of the solar energy available is absorbed by the plants, but the plant uses about half of this for its own metabolism, leaving only about 0.5% of the original amount of energy for the herbivores. The energy left over and stored represents growth.

Total energy taken in – Respiration = Growth

This is the energy available to the next trophic level. The herbivores then take their part, and so on up the food chain. As the available energy decreases, so must the number of organisms supported at each trophic level. This decrease can be illustrated with a *pyramid of numbers* and a *pyramid of energy* (Fig. 1-5). The pyramid of numbers shows that the organisms at the lower end of the food chain are the most abundant. Successive levels decrease rapidly in number until there are very few carnivores at the top. The pyramid of numbers ignores the total mass at each level—certainly one clover plant is not as important as one cow, but they count equally in this pyramid. On the other hand, the pyramid of energy indicates not only the amount of energy at each level but also the actual function of the various organisms in the transfer of energy.

Fig. 1-6 shows the yearly energy budget for a grass–meadow mouse–weasel food chain of a field ecosystem in Michigan. The ecosystem was much more complicated than that shown in the diagram, but for the sake of simplicity, other organisms such as plant-eating insects have been omitted. This example is also much simpler than most others would be, since the weasels feed primarily on the meadow mice rather than a large variety of animals.

Study the flow of energy and see where it is lost. A large amount is not available at the carnivore trophic level simply because it was not taken into the food chain at the producer and herbivore trophic levels. Only 1% of the solar energy available to the plants was utilized, and only 2% of the available plant material was consumed by the meadow mice. The unused plant material is left for the decomposers. Respiration at each trophic level accounts for loss of much of the energy. Of the total energy present at each trophic level, the following percentage is lost through respiration at that level: plants, 15%; mice, 68%; and weasels, 93%. There literally is almost nothing left for a higher trophic level (except the decomposers, of course). At the first trophic level 49,500,000 kilocalories* per year were available to the second level, but only about 400 kilocalories remain from the fourth level for a potential predator on weasels:

Weasel consumption – Weasel respiration = 400 kilocalories

It is obvious why there is no predator that feeds primarily on weasels. In summary, the most important point to be gained here is that energy does not follow a cycle. Because energy is lost as heat at each conversion, new high level energy must be constantly supplied from the sun.

BIOGEOCHEMICAL CYCLES

On Earth there is a finite amount of matter that does not increase or decrease. Living organisms require about 24 elements for normal growth. The most important of these are carbon, hydrogen, oxygen, phosphorus, potassium, nitrogen, and sulfur. Equally important but required in smaller quantities are calcium, iron, magnesium, boron, zinc, chlorine, molybdenum, cobalt, iodine, and fluorine. Since there is only so much matter available and since organisms must constantly take it in as food, it must be used over and over. For instance, matter is taken

*Kilocalorie = the amount of energy required to raise 1 kg of water 1° C. A pound of fat, for example, supplies about 4000 kilocalories. The average person consumes about 2600 kilocalories per day.

in as carbohydrate, fat, protein, vitamins, minerals, and a variety of other forms and is metabolized before being released as carbon dioxide, water, undigested material, and waste products. These materials flow from the nonliving environment into the living organism and then back to the nonliving environment in what are known as *biogeochemical cycles* (*bio* = living; *geo* = earth; *chemical* = the elements involved). If these materials did not move in this more-or-less circular path, all the available matter would soon be tied up in one part or another of the cycle, and none would be available for new living organisms. Because of this cycling, the elements can be passed round and round through the ecosystem. Some of these materials cycle very rapidly, others may remain for a short period of time in some part of the cycle, whereas still others may be tied up in the earth for millions of years.

However, all of these cycles can be classified into one of two types: *gaseous* or *sedimentary*. All those in the first category include a gaseous phase in some part of the cycle, and this tends to make the cycling much more rapid than it is in the nongaseous cycles. Although they vary from element to element, sedimentary cycles consist of a water-soluble phase and a rock phase. Both the gaseous and the sedimentary types of cycles are closely tied to another cycle—the water cycle. For this reason the water cycle, an example of a gaseous cycle, will be considered first.

Water cycle. Although most people may not be familiar with the ecological terms applied to the water cycle, they are nevertheless familiar with the events involved. Without the water cycle life would be possible only in the sea. As illustrated in Fig. 1-7, water does indeed travel in a cycle. Enormous amounts of solar energy are consumed in the evaporation of water (0.536 kilocalorie per gram). This water vapor is carried into the atmosphere, where it condenses and falls as rain or snow. Some of this falls on land. It may be evaporated back into the atmosphere, enter streams as surface runoff, or seep into the ground. Sooner or later all this water will return to the ocean, completing the cycle. Taking the earth as a whole, the water is cycling within a closed system. No water is lost, and none is gained.

It should be remembered that no other substance compares to water as a solvent. As water goes through its cycle, various substances dissolve in it and are carried along with the water. When the water evaporates, these dissolved substances are left behind. Water runoff from heavily fertilized agricultural fields is a potential source of pollution because of this fact.

Carbon cycle. The carbon cycle can be said to begin with photosynthesis, the primary pathway by which carbon is withdrawn from its gaseous state in the atmosphere and built into carbohydrate and other organic compounds. When herbivores eat the plants, the carbon is passed along in these organic compounds. Thus an important part of the carbon cycle is its movement from the carbon dioxide (CO_2) in air to plants and to animals further up the food chain. Carbon taken into a plant may travel a number of pathways. It can be metabolized by the plant or stored as protoplasm until the plant dies, at which time it serves as an energy source for the decomposers; or it can serve as food for some animal. As the energy is released at the various trophic levels, the carbon is again released into the air as CO_2. Not all the carbon built into molecules by living organisms is returned as CO_2. That part used in the formation of shells or skeletons may become incorporated into the Earth's crust and leave the carbon cycle for millions of years. Likewise, partially decomposed organic matter may accumulate and be changed into coal, oil, or gas, which also removes it from the cycle for long periods of time. When these fuels are burned, they again free carbon as CO_2 or carbon monoxide (CO) into the atmosphere.

An important aspect of the carbon cycle is that there are many pathways (Fig. 1-8) that a molecule can follow. If one pathway is somehow blocked, alternate pathways exist. The gaseous state of carbon also speeds up its distribution, making it a nearly "perfect" cycle.

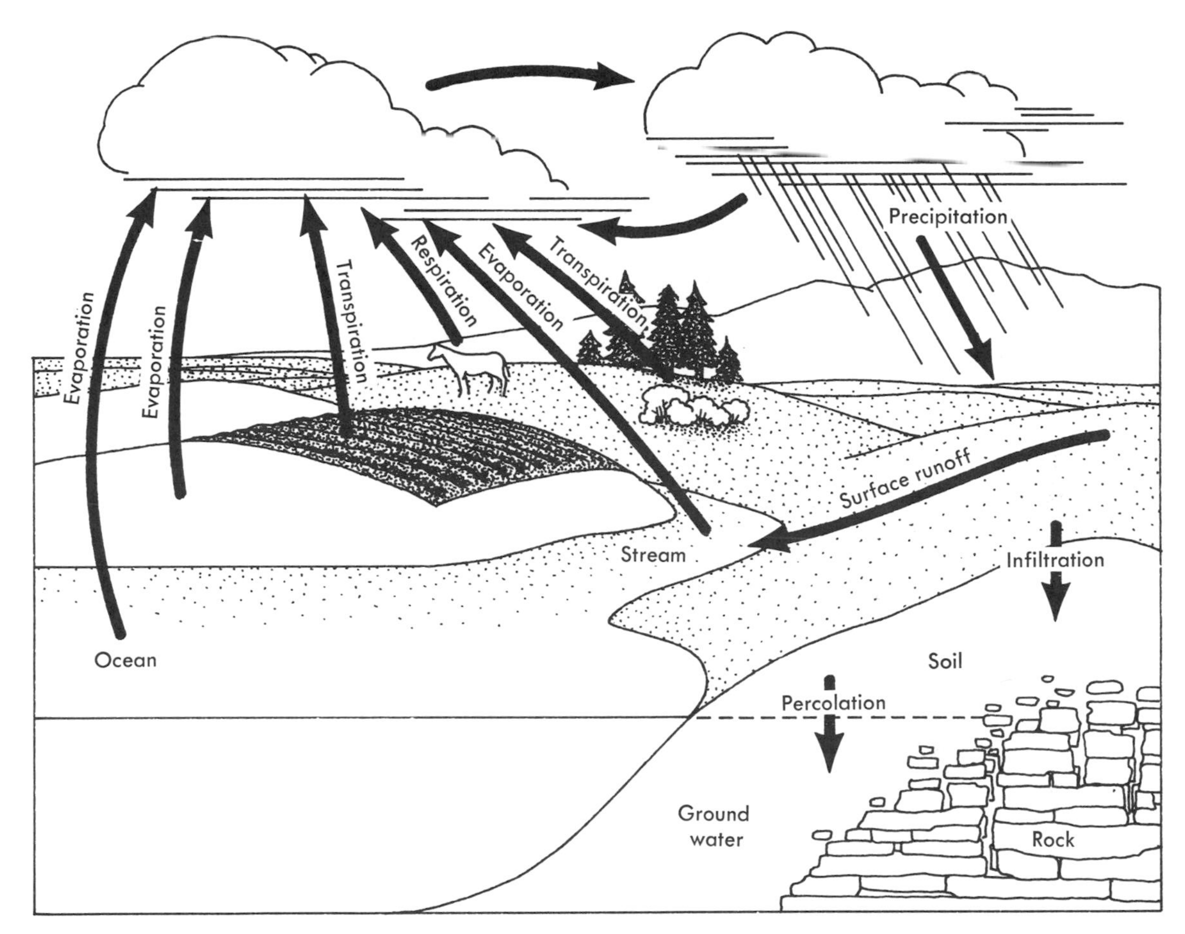

Fig. 1-7. The water cycle.

Nitrogen cycle. Although air is almost 80% nitrogen, the gaseous form (N_2) cannot be used directly by most organisms. Lightning causes some of the nitrogen of the atmosphere to react with other elements to form nitrates (Fig. 1-9), but this conversion process is of minor importance in relation to that of the *nitrogen-fixing* organisms.

The term *nitrogen-fixing* means to convert free nitrogen gas (N_2) into *nitrates* (NO_3^-). In modern agriculture, fertility of a field is improved by crop rotation involving legumes (beans and peas). The nodules on their roots are specialized structures inhabited by nitrogen-fixing bacteria. In water and in moist soil blue-green algae perform this same nitrogen-fixing function. These algae are important in maintaining fertility in rice paddies. The nitrates, whether formed by bacteria or algae, are absorbed by plants and incorporated into amino acids and proteins. The plant is either eaten by an animal or eventually dies. Either way, decomposers break down the proteins, leaving ammonia (NH_3). If the plant was eaten by an animal, its nitrogen would be excreted

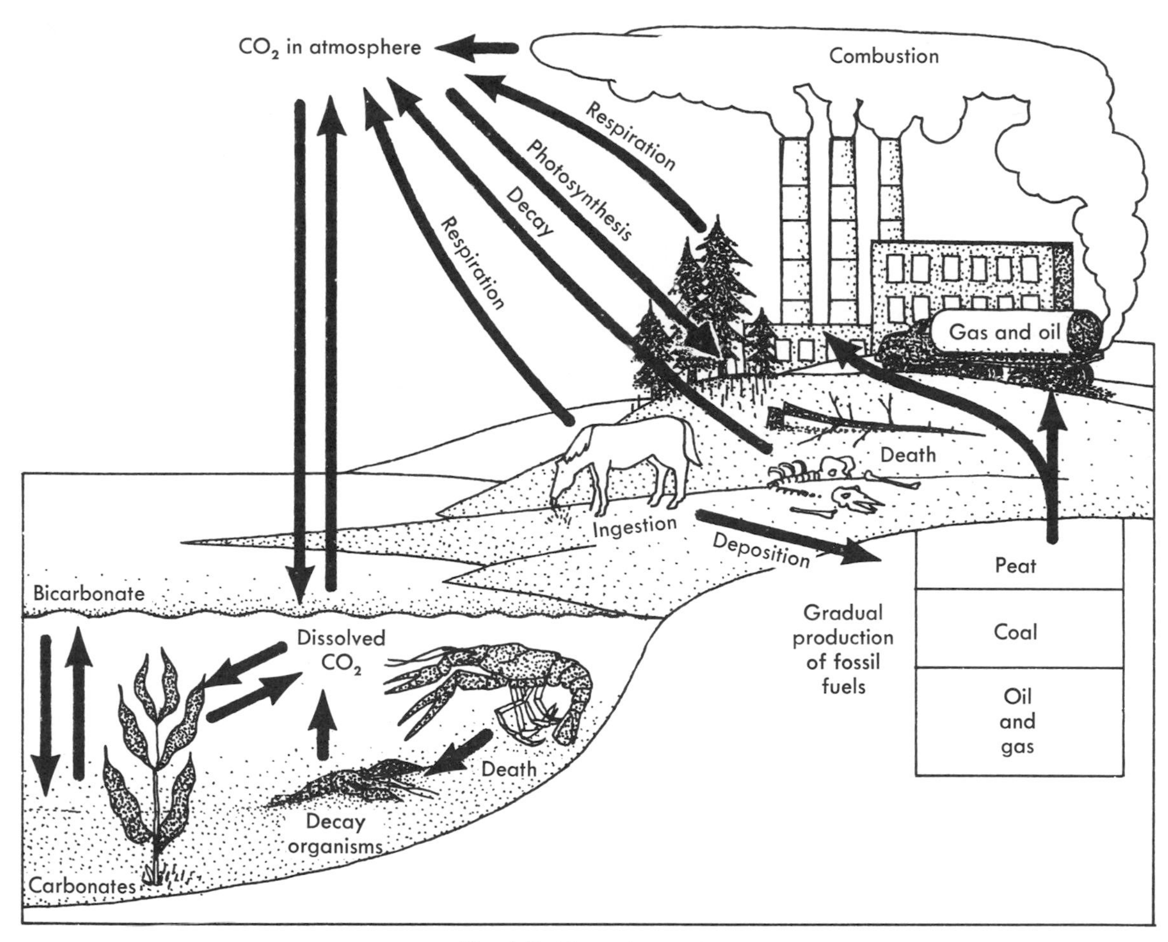

Fig. 1-8. The carbon cycle.

through the kidneys as a nitrogen waste product (usually ammonia, urea, or uric acid), but all of these will eventually be converted to ammonia. The ammonia is then utilized by *nitrite bacteria* to form nitrites (NO_2^-), which are in turn converted by *nitrate bacteria* into nitrates.

The nitrogen lost from the atmosphere is replaced by *denitrifying bacteria,* which use the nitrogen compounds as an energy source and release free nitrogen as waste. Some nitrogen is lost in runoff and may be tied up in deep-sea deposits, but this is balanced by "new" nitrogen released from deep within the earth by volcanic action. This is a relatively perfect, self-regulating cycle in which there is little overall change in the amount of nitrogen in each part of the world as a whole, despite rapid circulation of these materials.

Phosphorus cycle. In comparison to nitrogen,

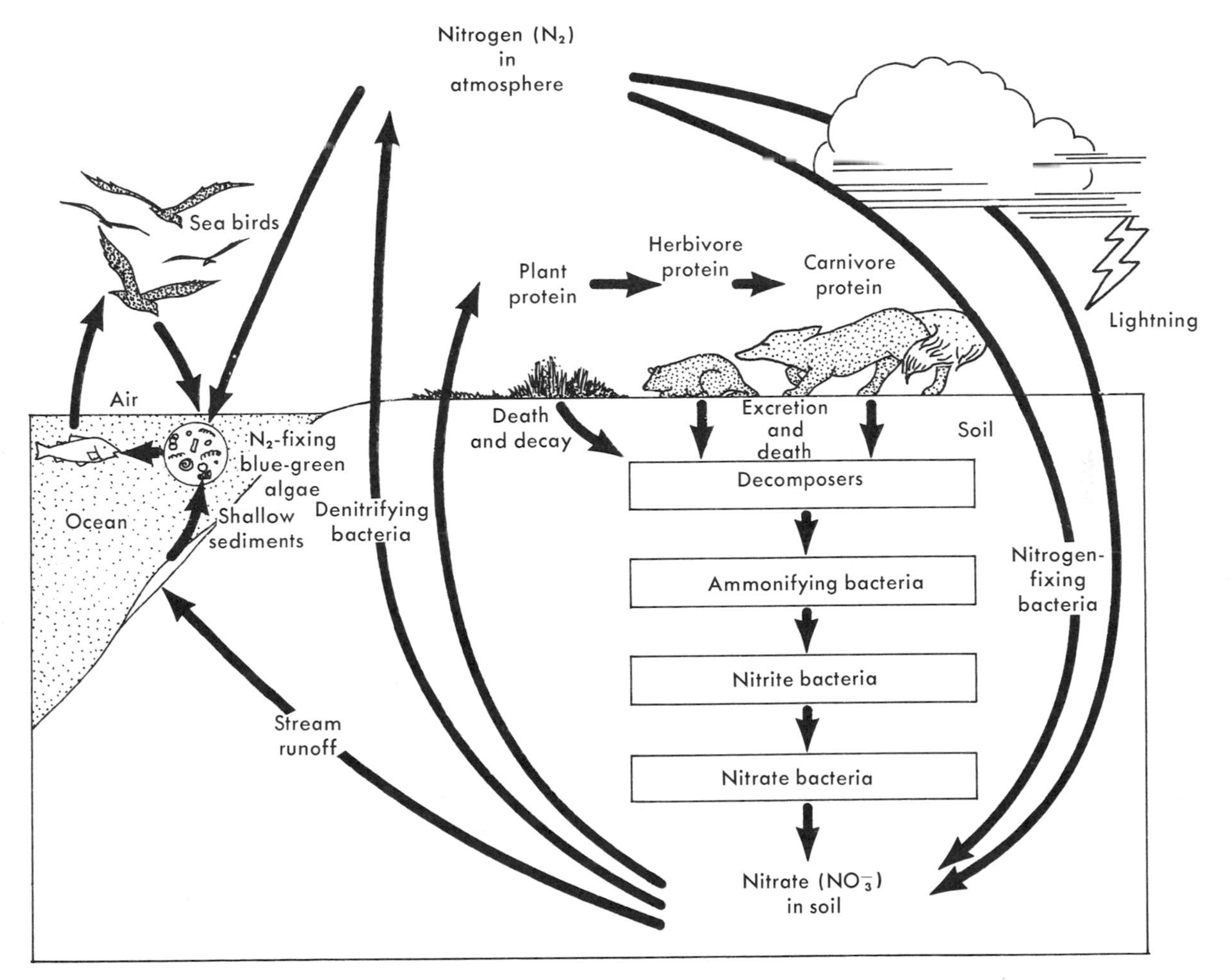

Fig. 1-9. The nitrogen cycle.

phosphorus (Fig. 1-10) is a rare material, but it is very important biologically. It is mined from apatite rocks and phosphate deposits and is used for agricultural fertilizer in the form of phosphate. Since it is soluble in water, it is carried away from the field into streams and lakes, eventually going to the sea. In the streams some of it is taken up by plankton and is cycled before it is eventually deposited at the bottom of a lake or ocean, where for all practical purposes it is lost from the ecosystem. Some phosphorus is returned by fishing, but the quantity is small and relatively unimportant. Fish-eating sea birds have created rich phosphorus deposits in a form known as guano (accumulated solid wastes).

Man has hastened the loss of phosphate by cultivation, and it is now estimated that we are losing 1 million to 2 million tons per year while returning only about 60,000 tons to the soil. There is no

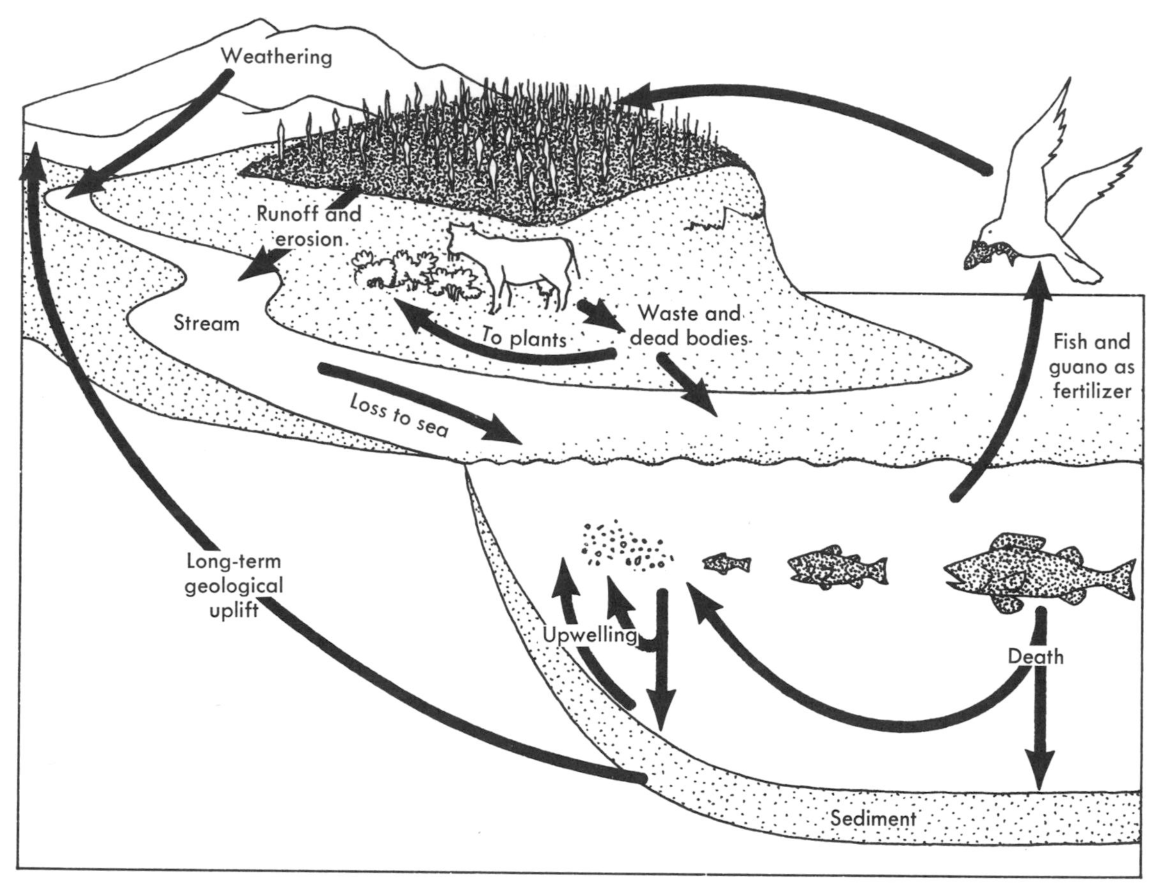

Fig. 1-10. The phosphorus cycle.

immediate problem, since we have vast phosphate deposits, but it does cause considerable damage to our rivers and lakes through algal blooms, which can kill fish and other organisms. Ultimately, however, the phosphate cycle could become critical through depletion of deposits or intolerable amounts in our waterways.

Notice that there is no gaseous phase in the phosphorus cycle. Also, the phosphorus reaches a certain place in its cycle and stops for long periods of time in a sedimentary state. This cycle is very definitely an imperfect one.

Other cycles. The foregoing examples illustrate the basic principles of material cycling. All materials cycle, some more efficiently than others of course, but they do cycle. Consider any of the known elements in the earth, or even man-made chemicals—they *all* cycle. When materials get in places where they are not wanted, we call this pollution. Many substances (mercury, DDT, and others) move rapidly

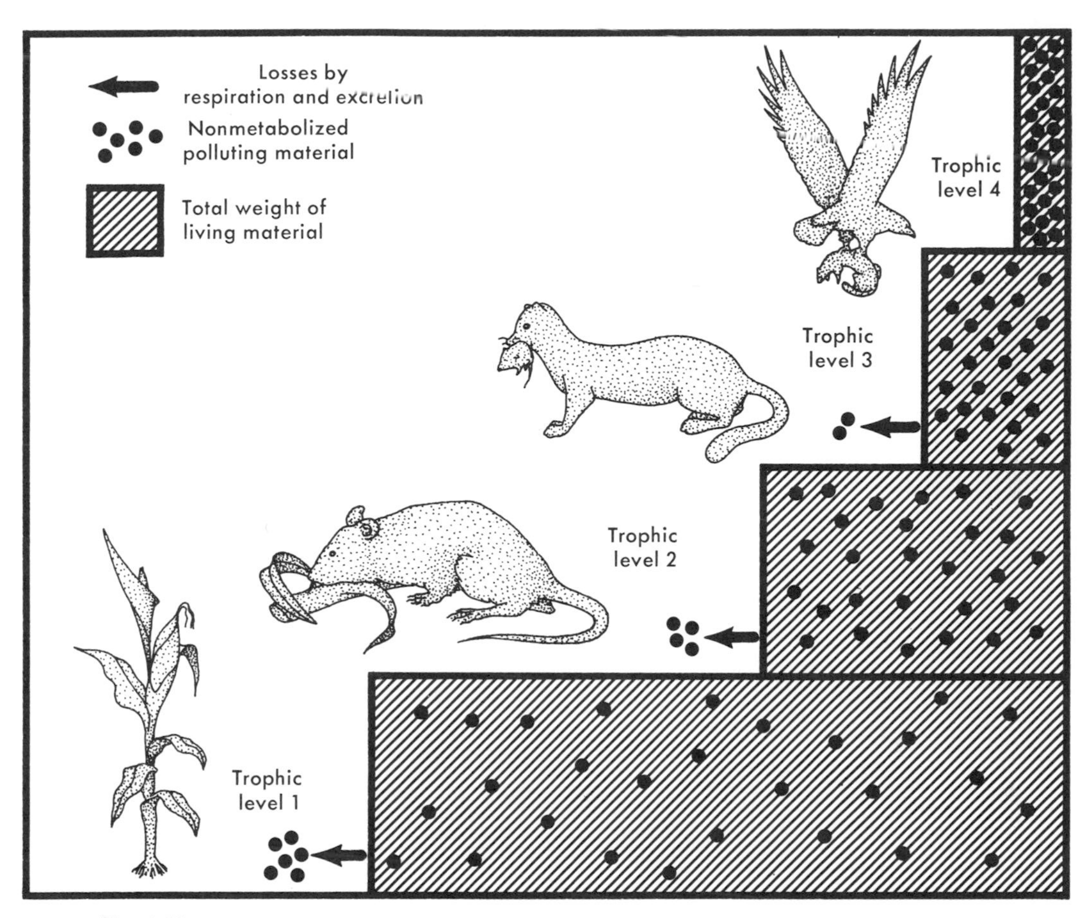

Fig. 1-11. Diagrammatic representation showing the increasing concentration of certain chemicals in the food chain. Large quantities of food material are taken in by each organism, but these are lost through respiration and excretion. If the chemical is not metabolized by the cells and remains there, the result is an ever-increasing concentration of the material as more food is taken in.

through the cycle to a certain point and then stop. They become concentrated and can create serious problems (Fig. 1-11). Other materials such as glass bottles, aluminum cans, or plastics will not cycle fast enough, and this also causes pollution problems. Biogeochemical cycles are essential to the functioning of ecosystems, but they can cause trouble when they do not work properly.

CELL STRUCTURE AND PHYSIOLOGY

Much of the material in the later chapters in this book will discuss the effects of a variety of environmental pollutants on the functioning of cells. In this chapter a few elementary concepts of cell structure and function will be introduced to allow a fuller appreciation of the specific action of these pollutants in the animal body.

CELL STRUCTURE

All complex organisms, plant and animal, are constructed of microscopic, living units called cells. Cells vary immensely in size. A bacterial cell may be less than 1 micrometer (μm) in length. (A micrometer is one thousandth of a millimeter; a millimeter is about 1/25 of an inch.) The cells of most animal tissues are about 20 μm in diameter, but these also vary greatly in size. Generally, a cell consists of a watery matrix called the *cytoplasm* and, suspended in this, a denser body called the nucleus (Fig. 2-1). Both the cytoplasm and the nucleus are bounded by a limiting envelope or membrane about 0.01 μm thick. The nucleus often contains a body called the *nucleolus*. Some cells produce structures external to the outer membrane. Plant cells are surrounded with a thick wall composed of a sugar (carbohydrate) called cellulose. Cells may also produce many short, hairlike structures known as *cilia* or a few longer structures called *flagella*.

Little structural detail can be seen in either the cytoplasm or the nucleus at the magnifications of the ordinary laboratory microscope. Under the electron microscope, which affords very high magnifications, the cytoplasm is seen to be a complex structure of membranes and small bodies, generally termed *organelles*. Among the more important organelles are *mitochondria,* which are usually oval or round and about 2 μm (or more) in length. Many hundreds are often present in each cell. The mitochondria are sites of certain energy reactions.

Coursing throughout the animal cell cytoplasm is a complicated array of membranes called the *endoplasmic reticulum* (ER). The membranes form flattened sacs and tubules. The ER is a transport channel for proteins and lipids that are distributed throughout the cell. The ER is also the site of synthesis of molecules known generally as steroids. Several hormones are steroids. Synthesized products within the ER can be pinched off, surrounded by ER membrane, and extruded out of the cell through the cytoplasmic membrane.

The ER membranes are connected with one or more areas of flattened, stacked membranes—the *Golgi bodies.* These latter membranes package proteins and carbohydrates (within their membranes) for export outside the cell. The Golgi membranes can also synthesize complex carbohydrates.

With the electron microscope it is seen that the ER of animal and plant cells is studded with small bodies known as *ribosomes.* These are about 0.02 μm in

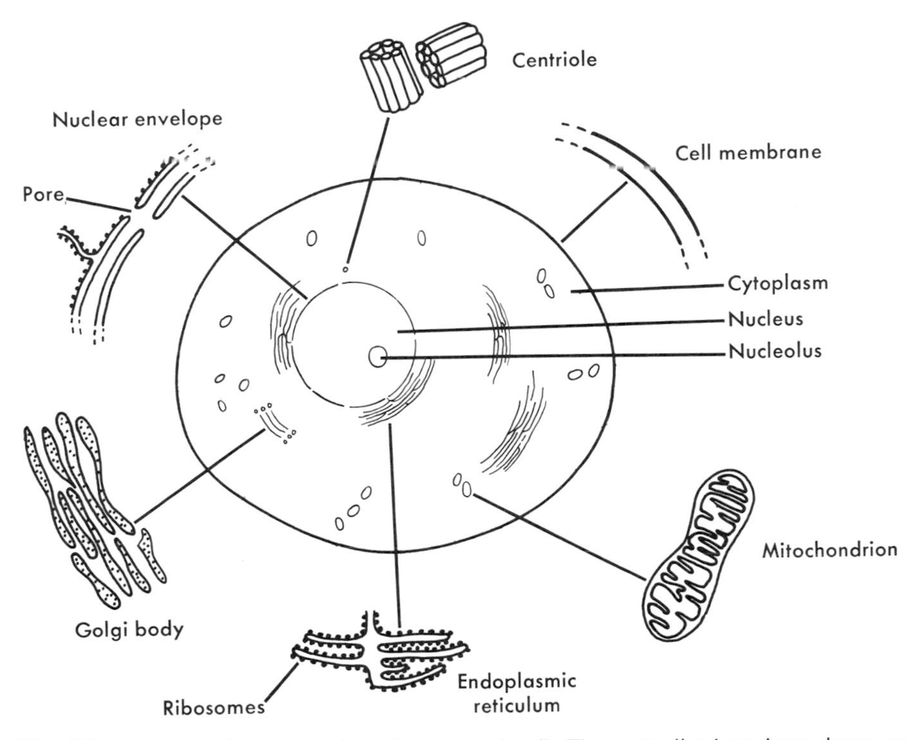

Fig. 2-1. Diagrammatic presentation of an animal cell. The organelles have been drawn as they might appear under the high magnification of the electron microscope. The centrioles function in cell division; the mitochondria are sites of certain energy reactions; the ribosomes and endoplasmic reticulum function in the synthesis of proteins; the Golgi body can "package" synthesized chemicals for export out of the cell.

diameter. The function of the ribosomes is singular: they are the sites of protein synthesis. Their composition is approximately half protein and half a phosphorus-containing large molecule, ribonucleic acid (RNA). There will be further discussion later about RNA and a similar molecule, deoxyribonucleic acid (DNA).

CELL FUNCTION

Reproduction

Reproduction by some form of division is one of the most significant characteristics of a living cell. Most cells (there are a few exceptions) *must* divide, that is, reproduce themselves, or they die. A division involves both the cytoplasm and its contents and the nucleus. The organelles of the cytoplasm are divided *approximately* equally between the two resulting (daughter) cells. The components of the nucleus, however, must be distributed *exactly* equally to the two new cells if these are to grow and function normally. The reason for this very fastidious requirement of nuclear division is simple: the nucleus contains the genes, that is, the units of hereditary information of the cell. Each cell resulting from a division must receive a *complete* copy of all this

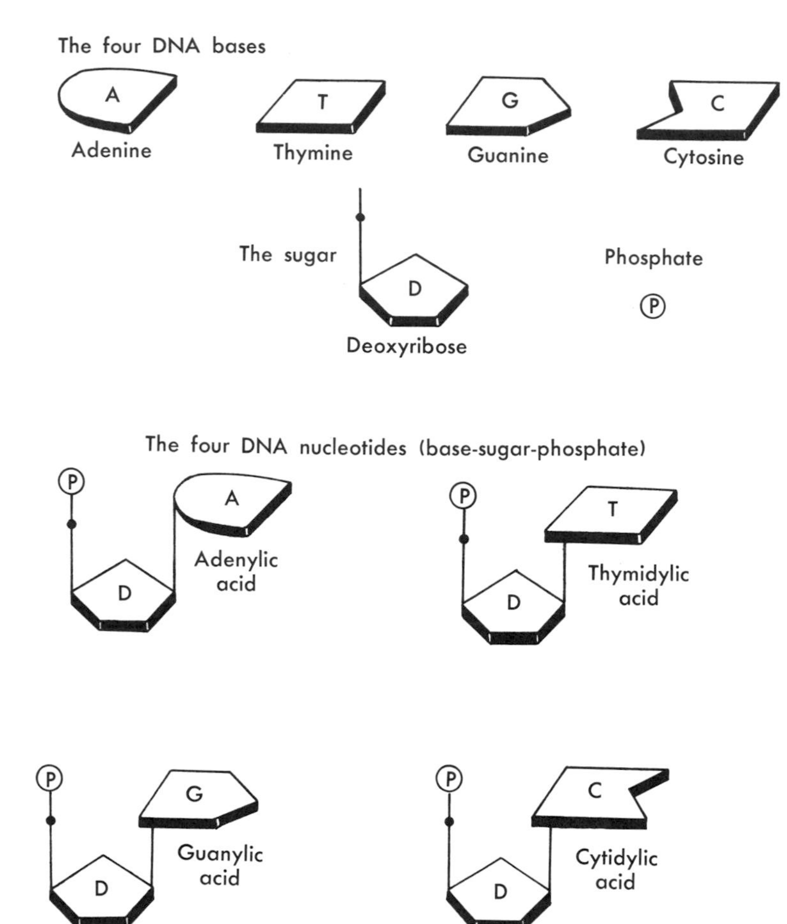

Fig. 2-2. Component molecules of DNA. The repeating unit of DNA is the nucleotide, which consists of one of four bases (adenine, thymine, guanine, or cytosine) plus a sugar (deoxyribose) and a phosphate (PO_3^{-2}). The base and the phosphate are bound to different places on the sugar. A DNA molecule consists of many nucleotides linked together by phosphates.

information, unaltered and unchanged, from the original (parent) cell. If some information is lacking (or indeed even in excess) in the new cell, it may malfunction or cease to function.

Structure of DNA. The genes consist of segments of a very large molecule of deoxyribonucleic acid (DNA). DNA is composed of many units called nucleotides, and thousands of these are bonded together to form a single DNA molecule. A gene consists of a few hundred to a few thousand nucleotides along the DNA structure. DNA itself consists of two chains of nucleotides wrapped around one another in a double helix. It is believed that only one of the helices functions in heredity.

The repeating unit of DNA, the nucleotide, is composed of a ring of carbons and nitrogens (the

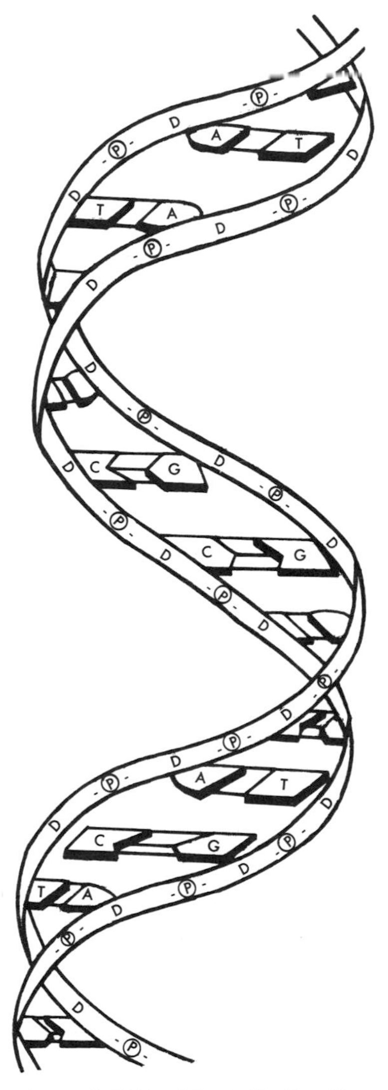

Fig. 2-3. Double helix structure of DNA. The helix "backbone" is made up of deoxyribose sugars (*D*) linked by phosphates Ⓟ. Pairs of bases are internal to the sugars and form the rungs of the helical ladder. The bonding of all these compounds is very specific. Note that the adenine (*A*) and thymine (*T*) of one helix are bonded, respectively, to the thymine and adenine of the other helix. The same is true for guanine (*G*) and cytosine (*C*) pairs.

base) joined to a phosphorylated, five-carbon sugar called deoxyribose. Several nucleotides are joined together by a phosphate bridge between the sugars of two adjacent nucleotides.

Four different bases may be attached to the sugar-phosphate so that four different nucleotide structures are possible. Figs. 2-2 and 2-3 are diagrammatic illustrations of these relationships.

The DNA of a nucleus is combined with protein to form structures called chromosomes. The chromosomes are visible only during cell division. The name means "colored body" and describes their appearance on being stained with certain dyes. Animal cells usually have many chromosomes and also a great deal of DNA. In animal cells the chromosomes always occur in morphologically identical pairs. Two identical chromosomes are said to be homologous to one another. The fruit fly, for example, has 4 pairs (in every cell nucleus), and the human has 46 pairs (in every cell nucleus).

Cell division. The critical requirement when a cell divides is that *both* new cells must each contain all the different chromosomes *in the paired state,* exactly as in the original parent cell. Barring mistakes, this means that the DNA is partitioned exactly equally and in an amount identical to that possessed by the parent cell. Division of the nucleus and equal distribution of its chromosomes is called *mitosis.* The entire cell cycle, from one mitosis to the next, can be illustrated most simply by a circular diagram (Fig. 2-4). The times noted are for a type of cell in culture. The actual division requires less than 45 minutes of the total 22 hours of the cycle. The G_1 phase is the active growth period of the cell, which leads to the S

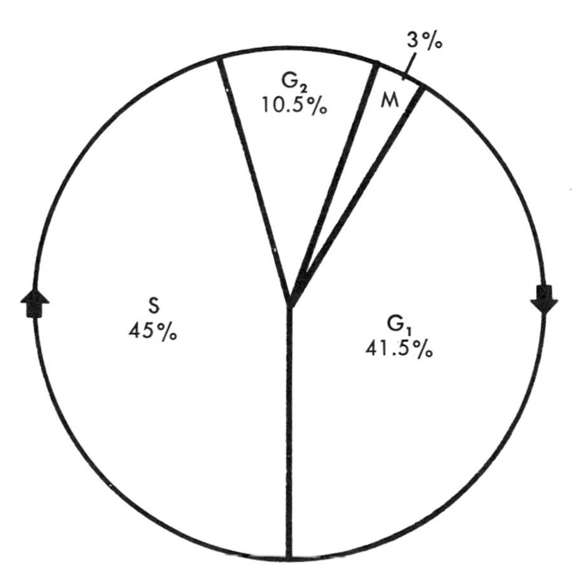

Fig. 2-4. The cell cycle. This diagram shows the various stages recognized in the life of a cell, from one division (mitosis, *M*) to the next. G_1 and G_2 are phases of cell growth and chemical synthesis, and *S* is the phase during which DNA is synthesized. The cycle diagrammed is based on a type of mouse cell in culture. The entire cycle requires 22 hours, and the percentage figures indicate that fraction of this time required by the different phases.

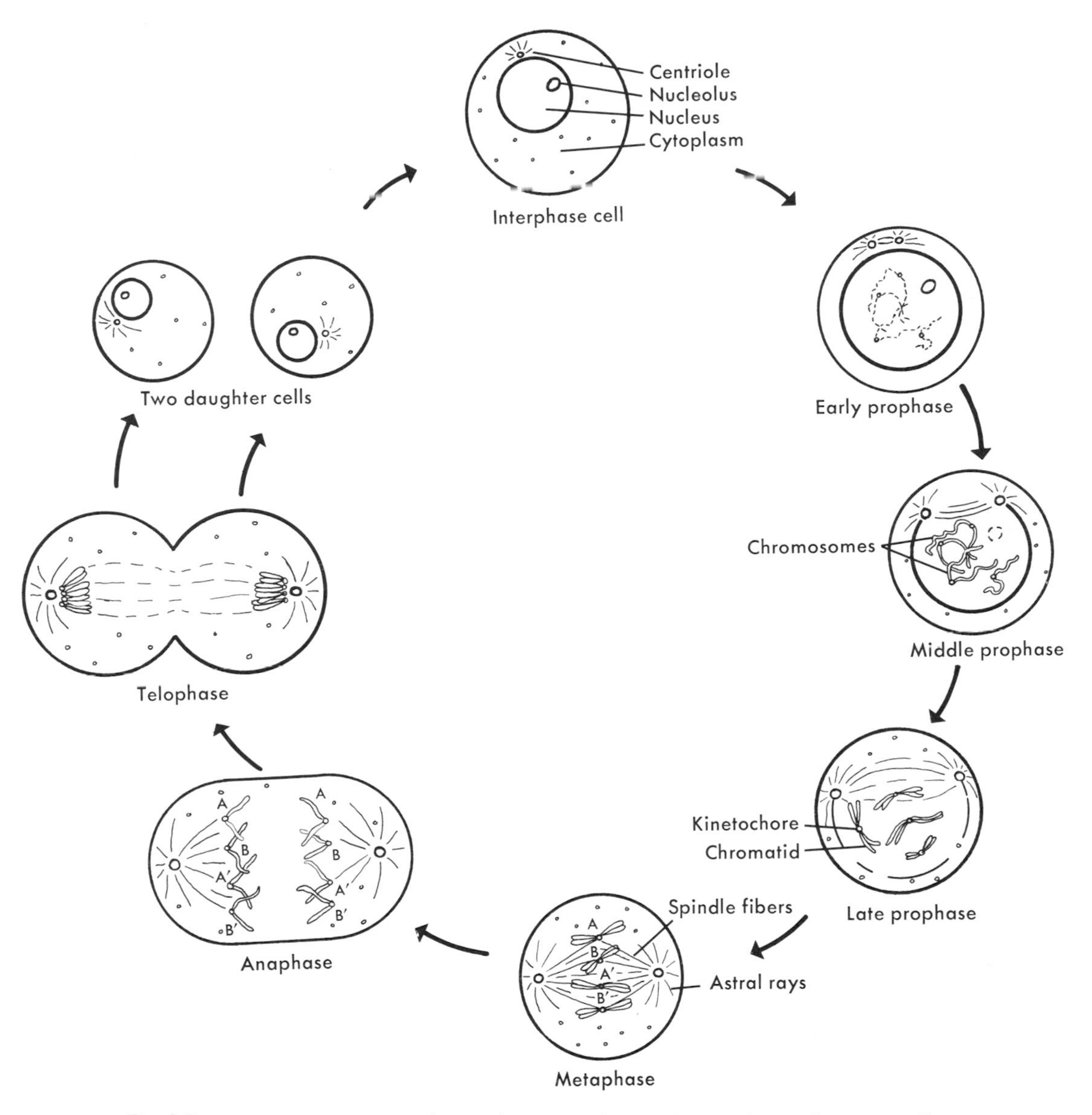

Fig. 2-5. Stages in the division (mitosis) of an animal cell. Interphase refers to a cell not undergoing division. The various prophases are characterized by the appearance of distinct chromosomes and disappearance of the nuclear membrane. During metaphase the chromosomes are aligned across the cell in a precise arrangement and are separated during anaphase. During telophase the original parent cell is separated into two daughter cells, each with identical numbers of chromosomes. Cell organelles are divided approximately equally between the two resulting cells. *AA'*, First homologous pair; *BB'*, second homologous pair.

phase, the period of DNA synthesis. During this phase (about 45% of the entire cycle), the DNA of the cell is doubled. The short G_2 phase is a division preparation stage where some proteins are probably synthesized. The following description of the actual mitotic events is illustrated by Fig. 2-5, which shows a cell nucleus with two pairs of homologous chromosomes. It is convenient to separate mitosis into four stages based on the activity and appearance of the chromosomes. These stages are *prophase, metaphase, anaphase,* and *telophase.* The G_1, S, and G_2 stages of the cell cycle are called collectively the *interphase* period.

Prophase. Strands of DNA and protein (the chromosomes) begin to appear in the nucleus as long filaments. Two small bodies, the *centrioles,* lying just outside the nuclear membrane, begin to separate from each other. Each will migrate 90 degrees, and in the process the nuclear membrane dissolves. From the centrioles arise many protein filaments, the spindle fibers, some of which will eventually attach to the chromosomes. Also radiating from the centrioles are other fibers, the astral rays. As prophase continues, the chromosomes shorten and thicken and their doubled nature now can be seen. Each chromosome is seen to be of two thick strands joined at a single point. This point is the *kinetochore.* Each of the two strands of one chromosome is a *chromatid.* By the time the centrioles have migrated to opposite sides of the cell, the nuclear membrane has completely dissolved and the nucleolus has disappeared.

Metaphase. After the events of prophase a (usually) brief stage, metaphase, ensues. This phase is characterized by a seemingly haphazard arrangement of the chromosomes near the center of the cell. Actually the arrangement is quite precise. The kinetochore of each chromosome lies exactly on an imaginary plate, the equator, that bisects the cell. Each kinetochore is joined by spindle fibers to both centrioles. Other spindle fibers appear to traverse the cell from centriole to centriole without interruption.

Anaphase. Anaphase stage is a relatively brief one. It is characterized by a movement of the sister chromatids (of each chromosome) apart from each other to opposite poles of the cell. Separation of chromatids requires a division or splitting of the kinetochore, which holds the sister chromatids together. Once separated, each chromatid now has its own kinetochore. The spindle fibers now appear to shorten, pulling the chromatids (now termed chromosomes) to the centriole at each pole. It should be noted that every chromosome sends one (and only one) chromatid to each pole. As these chromosomes near the centrioles, they become consolidated into an indistinct mass.

Telophase. The final stage of mitosis, telophase, is marked by a constriction of the cell along the old equatorial plate. As this "pinching in" progresses, the once distinct chromatids elongate and gradually become too thin to distinguish. The nuclear membrane and nucleolus reappear, and the two new daughter cells enter the G_1 phase of the next cell cycle. Later, in the S phase, the chromatids that entered the G_1 phase will duplicate and become double-stranded chromosomes in preparation for the next mitosis. Physical separation of the two new cells (a process called *cytokinesis*) completes mitosis. Through this precise distribution of chromosomes the daughter cells have received exactly the same quantity and quality of DNA as that possessed by the parent cell prior to its S phase. The cell organelles (mitochondria, ribosomes, and ER) are distributed approximately equally to the daughter cells. Ribosomes and ER are synthesized during the G_1 phase of the cell cycle; presumably the mitochondria, which contain some DNA, are also duplicated, but this event has not been observed.

The duplication of DNA during the S phase is an enzyme-catalyzed (see later), energy-requiring reaction. The double helix presumably opens up, exposing the bases of each nucleotide chain. Nucleotides in the nucleus may now pair with the exposed bases of each chain, according to certain bonding restrictions (adenine with thymine, guanine with cytosine). Completed pairing results in four chains (or two double helices after rewinding). One chain of each double helix is an original, the other is newly synthesized from the free nucleotides. Since mitotic chromosomes are made up of multiple DNA helices, the S phase must involve many DNA synthesis events.

Energy

Most cells can carry out all, or nearly all, of the functions that are usually ascribed to life: responsiveness, growth, metabolism (the sum total of myriad chemical and physical activities), and reproduction. One of the more important attributes of a living organism—and the cells which compose it—is that it transforms energy from one form to another. This attribute is in fact an absolute requirement. When energy transformation proves impossible, cells and organisms die. The ultimate source of energy, as pointed out in Chapter 1, is sunlight. Plants are able to trap this radiant form of energy via photosynthesis and use it to convert simple (inorganic) chemicals in the soil and atmosphere into complex (organic) chemicals required to maintain the living state. This is another illustration of the First Law of Thermodynamics, as stated in Chapter 1. The energy is not destroyed; it is merely transformed from one state to another.

In a true sense the green plant converts radiant energy of the sun into the structural energy of complex organic chemicals. This transformed energy can be viewed as being that of the bonds holding the atoms of the organic chemical together. This is not a strictly true description, but for the purposes here this view is sufficient. Both plants and animals are then capable of breaking the atomic bonds in such a way as to release and to utilize their energy to accomplish the work necessary for maintaining, repairing, and adding to their cellular machinery. At this point the concept of oxidation and reduction should be firmly grasped.

Oxidation and reduction. The process of *oxidation* is, by definition, the loss or removal of electrons. By the same definition reduction is the gain of electrons. When organic molecules are considered, the electrons most frequently lost or gained are parts of hydrogen atoms, each hydrogen atom consisting of one electron and one proton. Therefore, when an electron leaves or joins an organic compound, it usually carries a proton (hydrogen ion, H^+) along with it. The biologist, then, often thinks of oxidation and reduction in terms of the loss and gain of hydrogen *atoms*, but he must keep in mind that this is not a strictly correct definition because a proton does not always accompany a lost or gained electron.

It has been noted that respiration is the process by which glucose is completely oxidized to carbon dioxide and water. This means that all the hydrogen atoms are removed from the carbon atoms of the glucose, yielding the low energy form, carbon dioxide. The lost hydrogen atoms end up as part of water. At this point a cardinal rule of oxidation-reduction reactions must be noted: *whenever one molecule is oxidized, another must be reduced,* and vice versa. Neither electrons nor hydrogen atoms just leave a substance and float around free, although protons (hydrogen ions) may do so. Therefore one often speaks of one compound oxidizing or reducing another. It takes two.

The hydrogen atoms removed from glucose by cellular respiration ultimately reduce (add to) molecular oxygen to make water. That particular reaction, however, is the final step in the thirty steps or so that are involved in the whole process. Molecular oxygen does not react directly with glucose nor with any of its carbon-containing breakdown products. Instead, the hydrogen atoms are removed by, and temporarily bonded to, special molecules called coenzymes. Coenzymes serve as intermediaries, as hydrogen carriers, between the energy sources (such as glucose and the chemical intermediates of its breakdown) and molecular oxygen. Most coenzymes are vitamin derivatives; this function is, in fact, one of the most important reasons that vitamins are required by all living organisms. Perhaps the most important coenzyme in respiration is a derivative of the B vitamin niacin, called nicotinamide adenine dinucleotide (NAD). This and certain other coenzymes are also called *reducing agents* for their ability to readily give up electrons and protons to other compounds (thus reducing the latter).

Enzymes. These chemical reactions–the making and breaking of atomic bonds–in fact form the basis of life. However, they must take place at temperatures consistent with life, and these temperatures are relatively low (37° C for mammals, including the human species). At such low temperatures chemical reactions involving organic molecules occur very slowly. Unless they are speeded up, such reactions are useless to a living organism. The speeding up (catalyzing) of chemical reactions is accomplished by *enzymes.*

All enzymes are proteins. Most enzymes are very fastidious (i.e., specific) in the types of reactions they will catalyze. Some will speed the breakdown *or synthesis* of carbohydrates only; some are specific only for reactions involving other proteins; some are so specific that they catalyze the breakdown of *only one* kind of molecule. It is a truism that life on this planet would be absolutely impossible without cellular enzymes. Anything such as temperature, light (or the lack of it), or other chemicals that interferes with the function *or synthesis* of enzymes immediately places the life of a cell or organism in jeopardy. As will be discussed later, many of the chemical villains, or pollutants, do just that.

Energy reactions

The release of the energy represented by the bonds of complex molecules is a rather complicated business. The total process is called cellular respiration. There are many reactions involved (about thirty major steps), and some are not known in precise detail. These steps represent the basic pathway by which essentially all food material is broken down, oxidized in fact. Carbohydrates, such as starches and sugars, are broken down in the small intestine to smaller units (containing six carbon atoms) equivalent to glucose. Glucose can be considered as the starting material for these energy-yielding reactions of cellular respiration.

Proteins are degraded in the stomach to their component subunits, amino acids. Fats are similarly broken down in the small intestine to their subunits, for example, fatty acids. Amino acids and fatty acids, after certain preliminary reactions, enter into the overall pathway at various specific points. From then on they are oxidized via the same reactions as the carbohydrate material; in fact, the substances have become indistinguishable.

The overall reaction requires a sugar such as glucose (or a comparable compound) and oxygen and yields carbon dioxide and water. The following word equation shows this overall reaction:

$$\text{Sugar} + \text{Oxygen} \longrightarrow \text{Carbon dioxide} + \text{Water} + \text{Energy}$$

The sugar is a compound of carbon, as are all organic compounds, and, for glucose, has the formula $C_6H_{12}O_6$. The word equation may therefore be rewritten using the chemist's shorthand symbols for these compounds.

$$C_6H_{12}O_6 + 6O_2 \longrightarrow 6CO_2 + 6H_2O + \text{Energy}$$

This equation states that the energy in one molecule (unit) of glucose is released when "burned" with six molecules of oxygen; in the process six molecules each of carbon dioxide and water are liberated as "waste" products, and energy is made available to the cell. Beyond this, the equation tells nothing of how this reaction actually occurs.

ATP. None of the familiar energy sources (carbohydrates, proteins, fats) can directly donate their chemical energy to the myriad of different reactions which require that energy. Therefore there has evolved, apparently in the interest of economy, a common energy "currency"–adenosine triphosphate, commonly called ATP. The energy released from all these major sources is "captured" in the structure of this ubiquitous organic compound ATP. The ATP then serves in turn as the immediate energy donor for most of the energy-requiring reactions of living cells and organisms.

The structure of ATP can be symbolized by the following simple diagram:

ADENOSINE —Ⓟ~Ⓟ~Ⓟ

If the cell did not have the ability to "capture" the energy released in the burning (oxidation) of glucose, that energy would be liberated as relatively useless heat. Thus ATP is an important compound in the cell. Chemical reactions (the synthesis of enzymes and other proteins, for example) can usually only occur in the cell when one or more of the reactants is in a highly reactive state. ATP can readily transfer the energy of its structure to reacting molecules and so raise them to a higher energy condition. These high-energy molecules may then participate in a reaction.

ATP itself is formed from another compound called *adenosine diphosphate* (ADP) when an energy source is available. The energy is used to place a third phosphate group (a compound of phosphorus) on ADP as shown below. In these equations the symbol Ⓟ is used as a shorthand way of indicating a phosphate: HPO_4^{-2}. The last two equations are shown as reversible ($\leftrightharpoons$), since when ATP activates some other molecule, its energy is given up, its third phosphate (Ⓟ) is released, and ADP is the product. The squiggle bond between the Ⓟ's is used to indicate the unstable, and thus high-energy, condition created by the presence of the second and third Ⓟ's on ATP.

A more accurate representation of respiration, then, would be the following:

$$C_6H_{12}O_6 + 6O_2 + ADP + Ⓟ \rightarrow 6CO_2 + 6H_2O + ATP$$

Most of the energy that is made available in the oxidation of glucose is utilized to synthesize ATP. But no machine is 100% efficient, and the living cell is no exception. The cell's energy machinery is efficient enough to capture (as ATP) about 60% of the energy made available from the burning of glucose. The remaining 40% is lost as heat. This state of affairs is a direct result of the operation of the Second Law of Thermodynamics, which, it will be recalled, states that a decrease in the amount of useful energy occurs whenever an energy conversion takes place. This heat is not wasted in the sense that it is totally useless; it is this heat that maintains the body temperature of a mammal (including humans) at about 37° C.

The following concepts of cellular energy reactions have now been discussed:

1. Green plants transform radiant energy into chemical bond energy of complex organic molecules such as glucose. This sugar is a multicarbon molecule synthesized from carbon dioxide and water by photosynthesis. Such sugars can therefore be regarded as having radiant energy of sunlight stored in their chemical structure.
2. Animals and plants are capable of releasing the chemical energy stored in these sugar molecules. Complete release of this stored energy requires oxygen. These energy-releasing reactions involve the oxidation and degradation of the sugar and liberation of the original small

ADP + Ⓟ + Energy $\leftrightharpoons$ ATP

or

ADENOSINE —Ⓟ~Ⓟ + Ⓟ + Energy $\leftrightharpoons$ ADENOSINE —Ⓟ~Ⓟ~Ⓟ

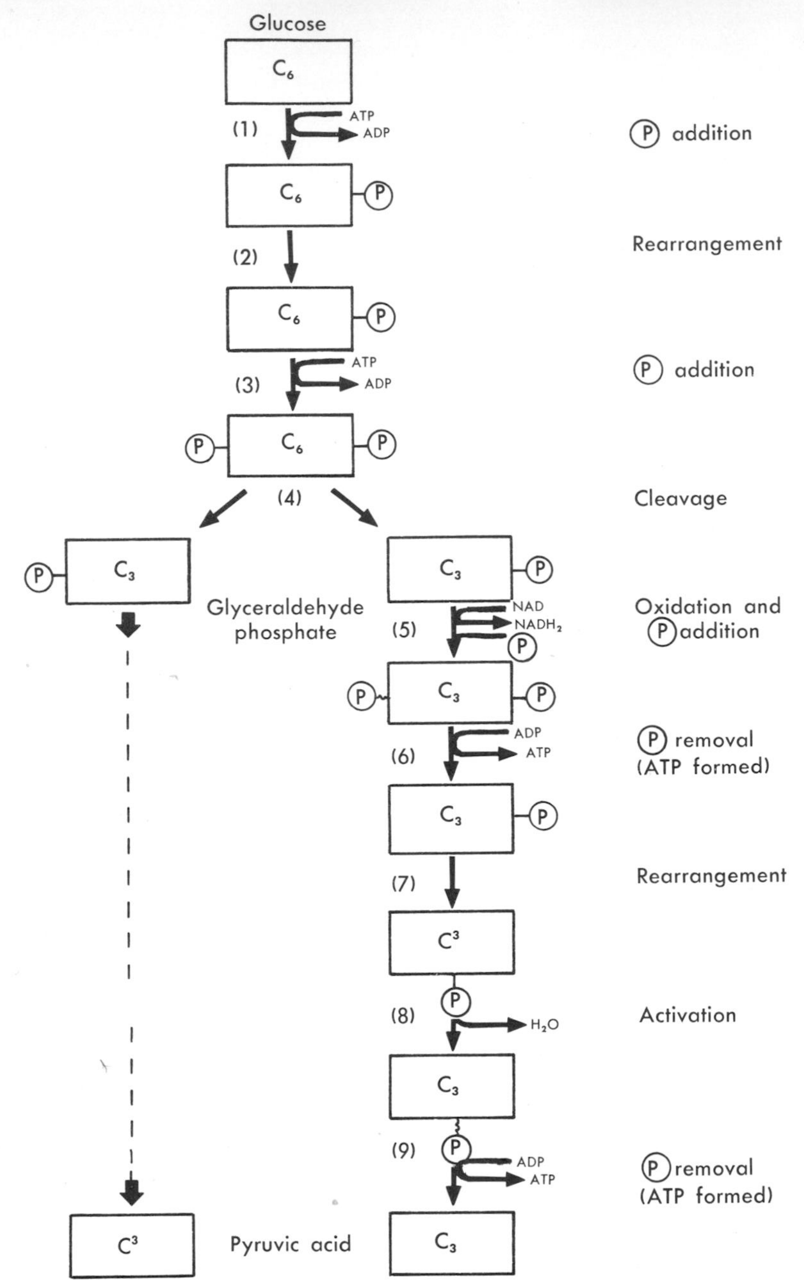

Fig. 2-6. Steps in glycolysis, or anaerobic oxidation. These steps result in the production of two molecules of pyruvic acid from the starting glucose molecule. Steps 1 through 3 are preparation steps for splitting the six-carbon glucose into two three-carbon molecules of glyceraldehyde phosphate (step 4). Steps 5 through 9 are primarily rearrangement steps leading to pyruvic acid. The vertical dashed line indicates that steps 5 through 9 occur for each glyceraldehyde phosphate. Two ATP's are consumed (steps 1 and 3), and a total of four ATP's are generated (steps 6 and 9). These steps therefore result in a *net gain* of two ATP's per glucose molecule. An $NADH_2$ is generated at step 5. These reactions take place in the cell cytoplasm.

molecules used in its synthesis (carbon dioxide and water).

3. Some of the energy released in the oxidation of glucose is lost as heat. Another portion is used to add a phosphate to ADP to form ATP. Since ATP functions as an energy "donor" in many energy-requiring reactions in the cell, its formation is the most significant step in these energy-release reactions.

The reactions themselves actually occur in three major blocks, each composed of many chemical steps. That the energy of glucose structure is released in small amounts by many steps represents a great deal of energetic wisdom: If all the stored energy were released in a single step, the cell would be burned to a crisp by the heat released from the resulting explosion. These three blocks of reactions are known as *glycolysis*, the *citric acid cycle*, and the *electron transport system*. Each individual reaction is catalyzed by a specific enzyme.

Glycolysis. As outlined in Fig. 2-6, glycolysis, or anaerobic oxidation, is the first block of reactions in the degradation of glucose. This sequence of chemical transformations eventually results in the breaking of each six-carbon sugar molecule into two three-carbon compounds, called pyruvic acid. Not only is pyruvic acid half the size of glucose but it is also more oxidized. Three reactions are required to activate and prepare the glucose molecule for the cleavage reaction. After cleavage the initial three-carbon substance (glyceraldehyde phosphate) is oxidized (that is, it loses a pair of hydrogen atoms [2H] to NAD) and eventually is converted to pyruvic acid. Fig. 2-6 is a simplified diagram of the glycolysis reactions.

Note that two molecules of ATP are used to activate the glucose to begin the sequence (reactions 1 and 3). After splitting (reaction 4), the three-carbon glyceraldehyde phosphate is oxidized by NAD and immediately picks up another phosphate group (reaction 5), this time in the form of free (inorganic) phosphate, not from ATP. The [2H] removed in reaction 5 comes from the carbon atom to which the new phosphate group is attached, and the oxidation of this carbon atom makes the carbon-phosphate arrangement very unstable. This new condition is represented by the squiggle in the product of reaction 5 (Fig. 2-6). In fact, the phosphate group would be more stable (less reactive) as part of ATP—even though, in comparison to *most* of the chemical compounds usually available, ATP is, of course, highly reactive.

In reaction 6 that unstable phosphate is transferred from the three-carbon glucose derivative to ADP, making ATP. In the next two reactions (7 and 8) the remaining phosphate group is shifted to another carbon atom, from which a molecule of water is then removed. This dehydration reaction creates another unstable situation, this time for the remaining phosphate group. So in the next reaction (9) the phosphate moves from the sugar derivative to ADP, forming another molecule of ATP, leaving the three-carbon compound, pyruvic acid, as the breakdown product of glucose.

Note that reactions 6 and 9 each produce one ATP. Actually, for every glucose molecule that goes through this process of glycolysis, reactions 5 through 9 occur twice. (This is indicated by the vertical dashed line.) Therefore, for each glucose, reactions 6 and 9 together produce *four* ATP's. The activation reactions (1 and 3) used two ATP's leaving a net gain of two molecules of ATP. Thus a summary equation for glycolysis would be as follows:

$$\underset{\text{Glucose}}{C_6H_{12}O_6} + 2ADP + 2\text{Ⓟ} + 2NAD \rightarrow \underset{\text{Pyruvic acid}}{2C_3H_4O_3} + 2ATP + 2NADH_2$$

No oxygen has been required up to this point. When oxygen is available, complete oxidation of the glucose carbons occurs and many more ATP's are formed (see later). However, when oxygen is not available, this is as far as cellular respiration goes. The end products are not carbon dioxide and water but, rather, a partially oxidized product called lactic acid, formed from pyruvic acid. Lactic acid accumulates in animal muscle, for example, during rapid contraction such as in running. Under these conditions ATP is being consumed by muscle cells very rapidly—too rapidly to be supplied even by the *complete* respiration of glucose because oxygen cannot be taken into the body and gotten to the muscle cells fast enough to support it. In other words, the physical inhalation and diffusion of oxygen are barriers.

However, the chemical reactions of glycolysis can speed up dramatically—fast enough, in fact, to supply the necessary ATP for rapid muscle contraction for some time. After the rate of ATP consumption subsides, the accumulated lactic acid is eventually removed (about 80% of it is reconverted to glucose—actually to glycogen, a compound of glucose). This reconversion is made possible by the ATP produced by the *complete* oxidation of the lactic acid by molecular oxygen and the remaining reactions of respiration.

Why does lactic acid accumulate and not pyruvic acid? There is only a little bit of the coenzyme NAD available; therefore it must be *recycled,* or else the whole process would be forced to a stop. In the absence of oxygen, the reduction of pyruvic acid to lactic acid *producing NAD* is the mechanism for recycling NAD so that glucose oxidation and thus ATP production can continue. When enough oxygen is available, it serves to recycle NAD (i.e., oxidize $NADH_2$ to NAD), thus allowing many more pairs of hydrogen atoms to be removed from the carbons of the original glucose molecule, resulting in its eventual complete oxidation to carbon dioxide (CO_2). More importantly, this whole process, described later, yields some thirty-odd more ATP molecules for each glucose molecule consumed.

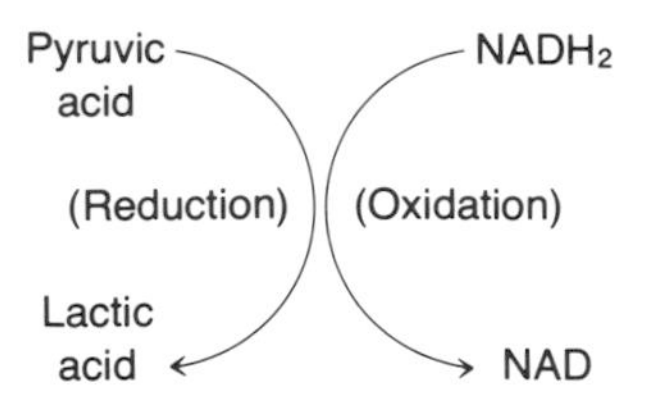

Citric acid cycle. *In the presence of oxygen,* however, the pyruvic acids are converted to a form that can gain entrance to the second major block of energy reactions, the citric acid cycle, or Krebs cycle. This conversion consists of a series of complex reactions, which need not be detailed here. Essentially, pyruvic acid has one carbon removed as carbon dioxide, and the remaining two-carbon fragment, the *acetyl* group, is attached to a molecule called coenzyme A (CoA). Removal of the CO_2 from pyruvic acid requires a reaction with a form of vitamin B_1 (thiamine). One step involves the generation of $NADH_2$.

$$\text{Pyruvic acid} + \text{Thiamine} + \text{NAD} + \text{CoA} \xrightarrow{(10)} CO_2 + \text{Thiamine} + NADH_2 + \text{Acetyl-CoA}$$

Since two pyruvic acids result from glycolysis, this reaction should be considered twice for each glucose. The acetyl-CoA represents the two carbons remaining from pyruvic acid decarboxylation (CO_2 removal) attached to coenzyme A. We should note that the two $NADH_2$ generated will be used later in the formation of ATP in the electron transport system.

The reactions in the citric acid cycle (and those of the electron transport system to follow) take place within cell organelles called mitochondria. A single liver cell may contain a thousand or more of these minute bodies. These reactions are cyclic in that the molecules are constantly regenerated. The cycle functions by accepting incoming molecules and systematically cleaving them to yield CO_2 and hydrogen atoms. The CO_2 is generally a waste product; it is brought to the lungs by the bloodstream and exhaled. The hydrogens, as they are produced, are bound to various coenzymes.

Rather than detail each step of the cycle, a general

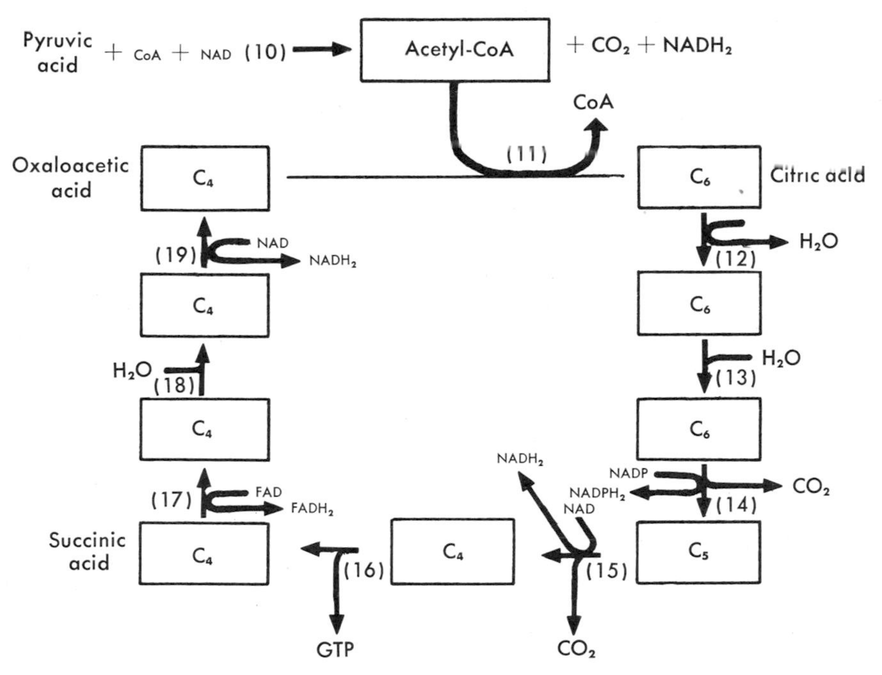

Fig. 2-7. The citric acid (Krebs) cycle. Removal of a carbon (as CO_2) from pyruvic acid (step 10) results in a two-carbon (acetyl) fragment. Attachment to coenzyme A (CoA) allows this fragment to enter the reactions. These two carbons are given off as CO_2 at steps 14 and 15, and oxidation of the intermediate compounds results in the production of several reducing agents ($NADPH_2$, $NADH_2$, $FADH_2$) at steps 14, 15, 17, and 19. A GTP (equivalent to ATP) is generated in step 16. The starting molecule of oxaloacetic acid is regenerated. The reducing agents carry electrons and protons and send these into the electron transport chain. The citric acid reactions take place in the mitochondria of cells.

description of the more significant reactions will be given. The reader should refer to Fig. 2-7 as these reactions are outlined. Since two acetyl-CoA molecules are produced in glycolysis from a single glucose molecule, these reactions must be taken twice.

The two-carbon acetyl group of acetyl-CoA is first added to a four-carbon acid (oxaloacetic acid), converting the latter to a six-carbon citric acid. Coenzyme A is released in this reaction (11).

Citric acid then undergoes two rearrangement reactions (12 and 13). The reactions involve the loss and then readdition of water.

The product of reaction 13 now undergoes the first of four oxidations in the citric acid cycle. Hydrogen atoms are removed, and one carbon is cleaved off as CO_2 (reaction 14). The coenzyme formed on accepting the hydrogens is $NADPH_2$, a molecule identical in structure to $NADH_2$, but with an additional phosphate group. The molecule resulting from this oxidation is a five-carbon acid.

The five-carbon acid now undergoes the second oxidation of the series and in two steps (reactions 15 and 16) is converted to the four-carbon succinic acid. These steps are complex ones and involve several molecules, including CoA, thiamine (vitamin B_1), and NAD. The products of these two reactions, in addition to succinic acid, are a CO_2, an $NADH_2$, and a molecule abbreviated GTP, which is guanosine triphosphate. Its structure is similar to that of ATP and in fact can be regarded as equivalent to the generation of an ATP molecule. The carbons of the two CO_2 molecules liberated up to this point may also be viewed as equivalent to the two carbons of the entering acetyl group. The $NADH_2$ formed by these steps carries hydrogen atoms (electrons and protons) that are released by the oxidation.

The four-carbon succinic acid now undergoes three reactions (17, 18, 19), the first and last of which are oxidations that generate additional coenzymes. The first oxidation results in the formation of the first of three remaining four-carbon acids and a coenzyme abbreviated $FADH_2$. This coenzyme (flavin adenine dinucleotide) has a structure somewhat different from $NADH_2$ or $NADPH_2$. The $FADH_2$ bonds the hydrogen atoms released by succinic acid oxidation (reaction 17).

The next two steps are a water addition reaction (18) and an oxidation reaction by NAD (19).

It will be noted that the final molecule in this cycle, oxaloacetic acid, is the starting molecule to which is added the acetyl group of acetyl-CoA. Since two acetyl-CoA molecules are produced by glycolysis, the reactions must be "run" twice to completely oxidize both acetyl groups. With the second "turn" of the cycle the six-carbon glucose that started glycolysis has been completely destroyed. Its carbons and oxygens have been released as six CO_2 molecules, and its hydrogen atoms are bound to a variety of coenzymes. Table 2-1 summarizes the major products of glycolysis and the citric acid cycle.

In these reactions emphasis has been placed on the metabolism of the sugar glucose and its breakdown product, the two-carbon acetyl group, in the citric acid cycle. This cycle is a central "hub" of all degradative reactions in the cell. Fats, which consist of long carbon chains, are oxidized by the cell, and the breakdown products are acetyl groups. These are combined with coenzyme A and so gain access to the citric acid cycle and are further degraded. Amino acids, the small molecules that bond together to form large protein molecules, may also enter the reactions at various points. Amino acids first have their amino group ($-NH_2$) removed, in addition to carbon side chains before entrance.

All classes of molecules then (sugars, fats, and proteins) are subject to degradation, yielding potential energy as hydrogen atoms bound to coenzymes, and the citric acid cycle is the single set of reactions responsible for all these degradations.

At this point in the cell's energy reactions only four ATP (including two equivalent GTP) molecules have been generated. These represent a capture of only a small fraction of the energy in the structure of the original glucose molecule. The rest of the energy is represented by the hydrogen atoms bound to the various reducing agents (coenzymes). Were it not for

TABLE 2-1. Summary of major products of glycolysis and the citric acid cycle*

	CO_2	$NADH_2$	$NADPH_2$	$FADH_2$	ATP (OR GTP)
From glycolysis	—	2 (5)	—	—	2 (6, 9)
From two pyruvic acid reactions	2 (10)	2 (10)	—	—	—
From citric acid cycle (2 "turns")	4 (14, 15)	4 (15, 19)	2 (14)	2 (17)	2 (16)

*Numbers in parentheses refer to reactions in Figs. 2-6 and 2-7.

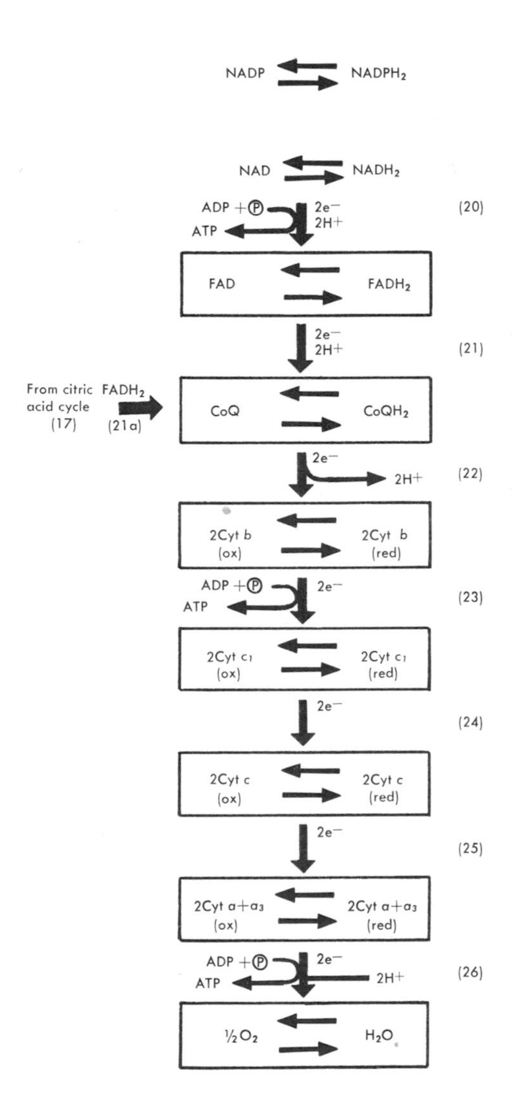

Fig. 2-8. The electron transport chain. The molecules in this chain accept and transport electrons carried by the reducing agents from glycolysis and the citric acid cycle. In their transport along the chain the electrons yield enough energy at three places (steps 20, 23, and 26) to cause the synthesis of ATP. The final electron acceptor at the end of the chain is oxygen, and the product is water. The oxidized and reduced forms of the component cytochromes are indicated by *(ox)* and *(red)*, respectively.

one additional series of reactions (the electron transport system), the energy in these reducing agents would be useless to the animal or plant cell. These highly active cells could not survive for long on the meager return of four ATP molecules per glucose molecule invested.

Electron transport system. The reactions comprising the electron transport system, or chain, also occur in mitochondria of animal and plant cells. The only two products of the chain are ATP and water (H_2O). The electrons and protons bound by the coenzymes from glycolysis and the citric acid reactions, are passed to molecules of the chain. At the end of the chain these electrons and protons combine with oxygen to produce water. Several (at least three) reactions in the chain release enough energy by electron passage to result in the synthesis of ATP from ADP and inorganic phosphate. As with the citric acid cycle, the electron transport chain reactions will be briefly described, referring to Fig. 2-8, which illustrates the steps.

There are known to be at least seven electron carriers in the transport chain. Most of these are proteins called *cytochromes.* The cytochromes have an atom of iron (Fe) as part of their structure, and it is the iron atom of each that is alternately reduced and then oxidized as it receives and gives up electrons respectively. On receiving electrons, the iron atom is reduced as follows: $Fe^{+3} + e^- \longrightarrow Fe^{+2}$. The electron is designated as e^-. Fe^{+3} is the oxidized state of the iron atom and Fe^{+2} is the reduced state. When this reduced atom gives up the electron to the next cytochrome in line, the donating iron atom returns to the oxidized state:

$$Fe^{+2} \longrightarrow e^- + Fe^{+3}$$

The first two carriers in the electron transport chain are not cytochromes. The first is the coenzyme FAD, which has already been seen in the citric acid reactions. The second is a noniron compound called coenzyme Q (CoQ). The electrons from both $NADH_2$ and $NADPH_2$ from glycolysis and the citric acid cycle are first passed to FAD of the chain. The FAD is reduced to $FADH_2$ (reaction 20). In this passage enough energy is released by the electron transfer to result in the phosphorylation of ADP to ATP. These electrons are now passed down the chain to CoQ and then to cytochromes b, c_1, c, and finally to a complex of enzymes designated a+a_3 (reactions 21 to 25). At each step the carriers are alternately reduced and oxidized as they receive and give up the electrons. At two additional steps in the chain enough of an energy change occurs to result in the synthesis of ATP. These steps are between cytochromes b and c_1 (reaction 23) and between the cytochromes a+a_3 complex and oxygen (reaction 26). The $FADH_2$ generated in the citric acid cycle reactions sends its hydrogen atoms along the same path, but they enter at the CoQ step (reaction 20a), reducing that molecule. Since the $FADH_2$ electrons (and protons) enter *after* the first ATP generating reaction, these electrons will result in ATP synthesis at only two steps ($b \longrightarrow c_1$ and a+$a_3 \longrightarrow$ oxygen).

The final electron transfer occurs when these electrons are removed from the a+a_3 complex and bound, with protons, to oxygen (reaction 26). Oxygen is reduced to form water in this last reaction. The oxygen that is the final electron and hydrogen acceptor in these reactions is atmospheric oxygen, inhaled by the animal. It is distributed to every cell of the animal body by the circulatory (blood) system (Chapter 3).

The mechanism by which ATP is synthesized in the electron transport chain is at present poorly understood. At least one protein enzyme is required (perhaps more).

With the formation of water by the chain the glucose that entered the glycolytic reactions is now completely oxidized to CO_2 and H_2O. Some of the energy of its chemical bonds has been conserved as ATP and some lost as heat. In these reactions it is seen that most of the ATP produced (about 87%) comes from the electron transport chain. It is possible to calculate how many ATP result from oxidation of a single molecule of glucose.

Reference to Table 2-2 shows a total of eight $NADH_2$ produced; each will result in the formation of three ATP in the chain. Each of the two $NADPH_2$ (from the citric acid cycle) will also generate three ATP. Each of the two $FADH_2$ will result in only two ATP, however. To these are then added the two ATP from glycolysis and the two (GTP) from the citric acid reactions. Table 2-2 summarizes these calculations, which result in thirty-eight ATP molecules per glucose molecule.

It was noted earlier that a lack of oxygen results in an altered reaction pathway for pyruvic acid. This molecule is reduced to lactic acid under such (anaerobic) conditions. It is now possible to understand the reasons why pyruvic acid cannot be prepared for entrance to the citric acid cycle under anaerobic conditions. Any circumstance that prevents electrons from being given up by the cytochrome a+a_3 complex (such as the absence of oxygen as an acceptor) will result in all molecules of the electron transport chain being permanently reduced, that is, loaded with electrons. If this happens, the coenzymes produced by the citric acid reactions cannot pass their electrons and hydrogens to the chain. Very quickly, all the reducing agents become loaded with electrons, further oxidations cannot occur in the citric acid cycle, and these reactions grind to a halt. Without a functioning citric acid cycle, pyruvic acid cannot be decarboxylated and so cannot enter the citric acid reactions. When this state of affairs exists, the pyruvic acid has only one route to take: it is reduced to lactic acid by oxidizing an $NADH_2$ to NAD. Most animal cells can survive only for short periods with only glycolysis operating. Without ATP from electron transport its minutes of life are numbered. In a real way this is an example of a holocoenotic environment (Fig. 1-1) at work inside the cell. A change in a single step of the energy reactions will affect many other steps in the reactions.

Protein synthesis

During the G_1 phase of the cell cycle (perhaps during the other phases also), the cell is actively metabolizing food molecules and generating ATP. One of the most significant energy-*requiring* reactions during this period is the synthesis of proteins. These large molecules not only form structural components (cell and organelle membranes, spindle fibers, ribosomes), but all *enzymes* in the cell are proteins. Virtually every chemical reaction in the cell is catalyzed (speeded up) by a specific enzyme. The enzymes present, and their amounts, dictate what kind of reactions will take place and how many in a given time.

Enzymes and other proteins are composed of many smaller molecules known as amino acids. (About twenty different amino acids occur commonly in nature.) The following is a generalized structure of the amino acids:

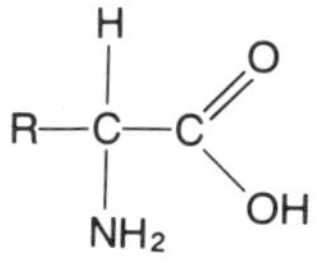

$$R-\underset{NH_2}{\overset{H}{\underset{|}{\overset{|}{C}}}}-C\begin{matrix}\nearrow O \\ \searrow OH\end{matrix}$$

TABLE 2-2. ATP production by reductants of the energy reactions; electrons of $NADH_2$ and $NADPH_2$ result in three ATP each; those of $FADH_2$ result in two ATP each

REACTIONS	REACTION NUMBERS	REDUCING AGENTS (COENZYMES)	ATP FROM ELECTRON TRANSPORT
Glycolysis	(5)	2 $NADH_2$	6
Pyruvic acid decarboxylation	(10)	2 $NADH_2$	6
Citric acid cycle (2 "turns")	(15, 19)	4 $NADH_2$	12
	(14)	2 $NADPH_2$	6
	(17)	2 $FADH_2$	4
ATP from glycolysis (net)	(6, 9)		2
ATP (as GTP) from citric acid cycle (2 "turns")	(16)		2
		Total ATP	38

In this structural formula the $-NH_2$ is the amino group, $-C{\lesssim}^{O}_{OH}$ is the acid (carboxyl) group, and R represents one of twenty different side groups, which may be a hydrogen atom, a carbon chain, or a ring structure. Proteins differ from one another in the *number* of amino acids bonded together (from hundreds to thousands) and in the *order* of amino acids bonded.

The specificity of an enzyme in catalyzing one type of reaction and not another lies in its three-dimensional shape, and this shape is determined in large part by the sequence (order) of its constituent amino acids. What types of chemical reactions occur, how many, and when, determine not only the physiological capabilities and morphology of the cell but also those of the organism of which it is a part. What an organism does and will become therefore depends on its enzymes; these in turn vary with their amino acid sequence.

Heredity consists of the passing of information from one generation to the next–information that determines the physical similarity between parent and offspring. This information consists of specific instructions on what kind and how many enzymes the offspring may synthesize. More specifically, it is a set of instructions that dictate amino acid sequence for each of the thousands of enzymes required. The information molecule that contains these instructions is DNA. DNA is the only molecule of major significance that is passed from male to female during reproduction. The way in which DNA functions in protein synthesis in the cell now will be examined.

The information carried by DNA lies in the sequence of bases making up its nucleotide chains (Figs. 2-2 and 2-3). This base sequence is eventually translated into the amino acid sequence of a protein (enzyme). DNA is almost wholly restricted to the cell nucleus. Protein synthesis, however, takes place in the cytoplasm, at the ribosomes specifically. Since DNA does not leave the nucleus, its information must be carried to the site of protein synthesis. This information transfer is conducted by another molecule, which is a type of RNA and is specifically known as messenger RNA (m-RNA). RNA is similar in general structure to DNA, but there are several differences. RNA uses the sugar *ribose* instead of the *deoxyribose* of DNA, and the base *uracil* replaces the closely similar *thymine.* The other three bases are the same as those found in DNA. Also, RNA molecules are single chains, not double helices as in DNA.

Messenger RNA is enzymatically synthesized in the nucleus. Its base sequence is determined by the base sequence of that segment of DNA involved. Free RNA nucleotides (base-sugar-phosphate) in the nucleus are base-paired with corresponding nucleotides of DNA, which unwinds, exposing its bases. The same pairing restrictions hold for RNA synthesis as for DNA synthesis: DNA guanine with RNA cytosine (and DNA cytosine with RNA guanine), and DNA thymine with RNA adenine (and DNA adenine with RNA uracil). Although the RNA bases are bonded to the DNA bases, the union is only a temporary one. As soon as the RNA base sequence is determined and their nucleotides are phosphate-bonded to form a chain, the m-RNA molecule comes off the DNA pattern, leaves the nucleus, and binds to a ribosome in the cytoplasm. The m-RNA is now carrying the information present in the base sequence of DNA. Fig. 2-9 diagrams m-RNA synthesis. These reactions have been termed *transcription.*

Numerous experiments have shown that the m-RNA base sequence determines the amino acid sequence, and so the type of protein to be synthesized. Furthermore, a group of three contiguous bases on m-RNA (the codon, so called) is required to specify the position of one amino acid. Although base order specifies amino acid order, there is no known mechanism that allows an amino acid to bond directly to a m-RNA base. Instead, the cell interposes another type of RNA as an *amino acid carrier* molecule, called transfer RNA (t-RNA), which binds to free amino acids in the cytoplasm and carries these to a specific base triplet (the codon) of m-RNA. The

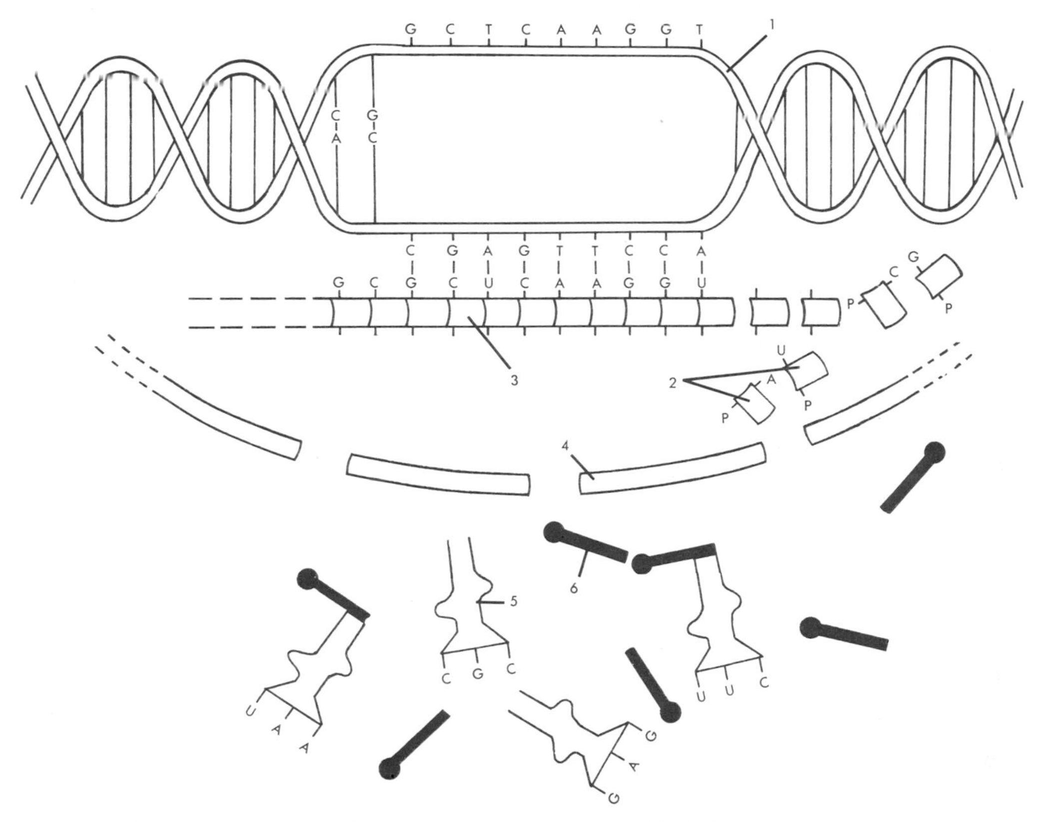

Fig. 2-9. Transcription steps of protein synthesis. The double helix DNA unwinds to expose unpaired bases. RNA nucleotides then base-pair with the DNA bases to form a long molecule of messenger RNA. Transfer RNA is also synthesized by the transcription process. In animal and plant cells these reactions take place in the nucleus. Both types of RNA ordinarily leave the nucleus and function in the cytoplasm. Transfer RNA attaches to and carries amino acids, and messenger RNA attaches to the cell's ribosomes. *1*, DNA; *2*, RNA nucleotides; *3*, messenger RNA; *4*, nuclear membrane; *5*, transfer RNA; *6*, amino acids.

t-RNA–amino acid binding reaction is enzyme catalyzed and requires ATP. There is at least one specific t-RNA for each different amino acid (in fact, a specific amino acid may bind to several different t-RNA molecules). The t-RNA's are relatively small molecules of about 80 nucleotides long. They are transcribed, like m-RNA, from a DNA segment in the nucleus.

On each t-RNA there occur at at least one position three exposed (unpaired) bases. This triplet is called the anticodon. This t-RNA triplet will base-pair with a m-RNA codon triplet, and the pairing will be specific according to the same restrictions governing m-RNA transcription; that is, a codon on m-RNA of the base sequence AUG will bind specifically to a t-RNA having the anticodon UAC. The codon AUG specifies the amino acid *methionine*, and this will be the amino acid carried to this m-RNA codon position

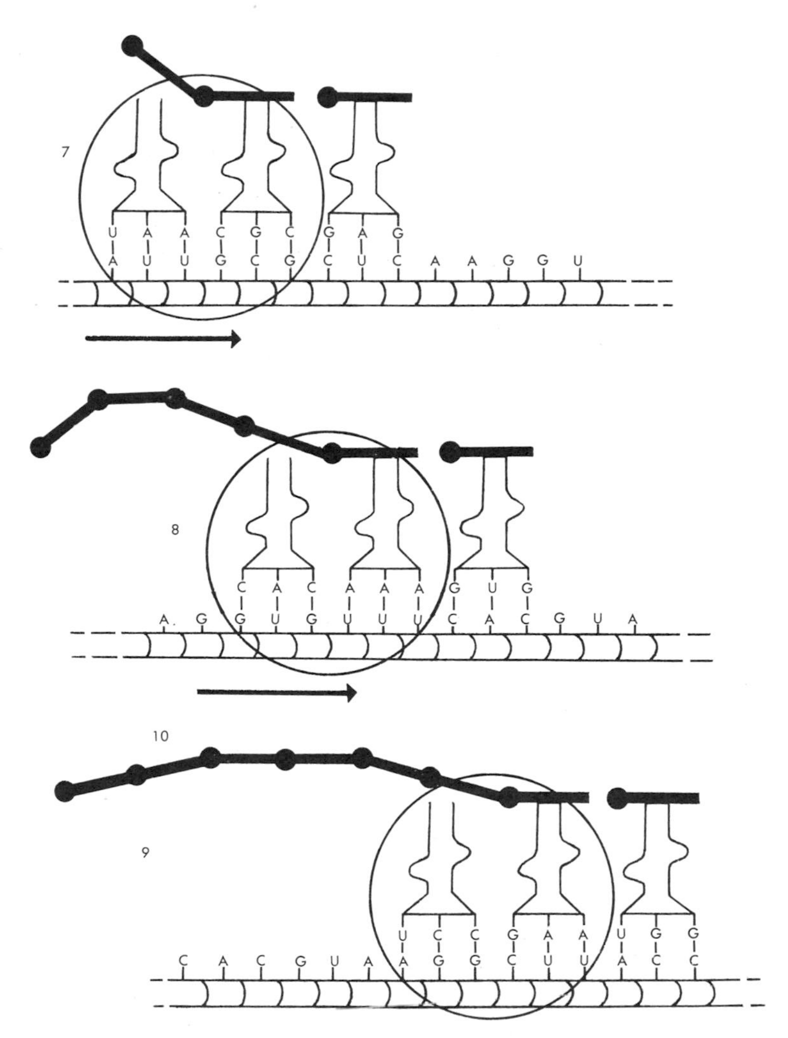

Fig. 2-10. Translation steps in protein synthesis. The base sequence of messenger RNA specifies the sequence of transfer RNA and therefore amino acid sequence. Transfer RNA leaves the ribosome on donating its amino acid to the next transfer RNA in line. Amino acids are linked together only at the ribosome; therefore the ribosome must move along the messenger RNA for the latter's bases to be "read out" by the transfer RNA's. Complete reading of messenger RNA results in a string of amino acids linked together, that is, a protein. *7*, Ribosome; *8* and *9*, steps in translation; *10*, growing amino acid chain (protein).

by a t-RNA. If the second m-RNA codon is UGG, it will pair specifically with a t-RNA having the anticodon sequence of ACC. UGG specifies a t-RNA carrying the amino acid *tryptophan.*

As each new amino acid is brought to its specific position, it is enzymatically bound to the preceding amino acid (the enzyme is one of the ribosome proteins). GTP is required as an energy source (acting like ATP) for amino acid union. The first amino acid, on binding to the second, is freed from attachment to its t-RNA. This t-RNA then leaves the m-RNA codon to "pick up" another specific amino acid. By "reading out" the m-RNA codons, t-RNA's line up their amino acids in a specific sequence, dictated by the sequence of codons. These events at the ribosome are called *translation* and are illustrated in Fig. 2-10.

Translation of m-RNA codons continues along the entire length of the m-RNA, each new amino acid being bound to the preceding one. By the time the last codon is translated, the final t-RNA has bound to it a long string of amino acids. These form the newly synthesized protein. The final codon on a m-RNA is one of the three known "stop" codons. It is a signal that translation is over, that the entire protein has been synthesized. No more t-RNA's are added, and the new protein is freed from its attachment to the final t-RNA that was translated. Since a codon of three nucleotides (bases) is required to position one amino acid, a m-RNA 1500 bases long will translate a protein of 500 amino acids. Since proteins are very variable in size, m-RNA's exist that are both shorter and longer than 1500 nucleotides.

In the final analysis, DNA is the important molecule in all these reactions; DNA is the master blueprint. It alone carries the information on what enzymes (proteins) will be synthesized and allows this information to be used by m-RNA and t-RNA. m-RNA and t-RNA are expendable; they are constantly replaced by transcription in the nucleus. DNA, however, is not expendable; DNA is the only information that can bridge the generation gap.

3 PHYSIOLOGY

Although humans and other animals are put together in different ways, they all face essentially the same problems—obtaining food, digesting and releasing energy from food, removing waste products, obtaining oxygen, coordinating body function, and so on. Although the methods used to solve these problems vary from one type of animal to another, the basic mechanisms are remarkably similar. This similarity is of great importance in determining the effects of a particular substance on an individual species: if a substance is harmful to one species, it will very likely cause similar effects in other species.

In this chapter some basic physiological functions are discussed. Although the emphasis will be on human mechanisms, it should be kept in mind that these same mechanisms exist in most other organisms.

GAS EXCHANGE

The discussion of cellular metabolism in Chapter 2 pointed out that there are two major metabolic processes in animal cells—*anaerobic oxidation* (called glycolysis), which requires no oxygen and does not produce carbon dioxide, and *aerobic oxidation* (involving the citric acid cycle and the electron transport system), which requires large quantities of oxygen and produces large quantities of carbon dioxide. To supply oxygen and remove the carbon dioxide most higher animals have developed a mechanism for gas exchange called the *respiratory system*.

Anatomy of a lung. In a typical vertebrate such as a frog, alligator, or human, atmospheric air is exchanged with the body tissues through a pair of lungs. A lung (Fig. 3-1) is a chamber lined by moist epithelium and supplied with a network of blood capillaries. Air passing to the lungs goes (1) through the mouth or nose; (2) through the *glottis*, an opening in the floor of the pharynx; (3) into a cartilaginous framework, the *larynx* (voicebox); (4) through a *trachea* (windpipe), which extends into the thorax (chest) and branches into (5) two *bronchi*, one to each lung; the bronchi, in turn, branch into much smaller *bronchioles*, which finally terminate in (6) microscopic compartments called *alveoli*.

The alveoli are in close communication with the blood, having only two layers of cells separating the air from the red blood cells. To reach the red blood cells oxygen in the gas phase in the alveolus crosses the alveolar cell membrane, then crosses the blood capillary membrane, diffuses through the liquid portion of the blood, and passes into the red blood cell (Fig. 3-2). Each human lung contains an estimated 700 million alveoli with a total surface area of approximately 70 m^2. The capillaries covering these alveoli contain only about 100 ml of blood at any given time so that it must be spread very thinly, making gas exchange easier. This situation is analogous to spreading half a glass of water evenly over the floor of an average-sized classroom. The alveoli of mammalian lungs, including those of man, are bound

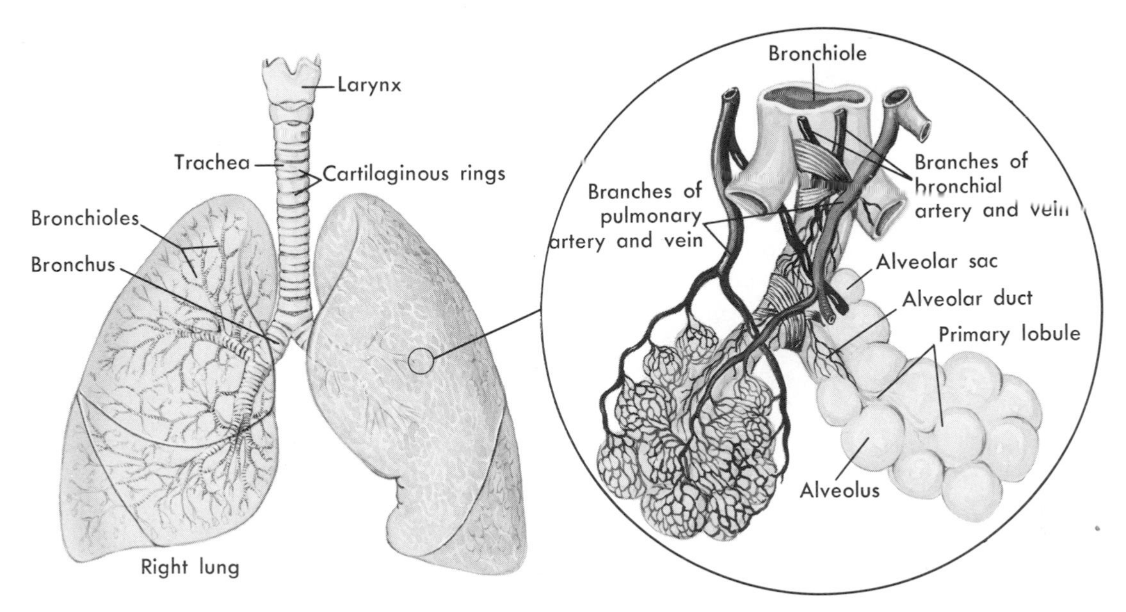

Fig. 3-1. Respiratory system. Human lungs with the right lung cut away to show the bronchi and bronchioles. **Inset,** Alveoli, or terminal air sacs, in the lung and the relationship of the blood supply to these structures. (Modified from Schottelius, B. A., and Schottelius, D. D.: Textbook of physiology, ed. 17, St. Louis, 1973, The C. V. Mosby Co.)

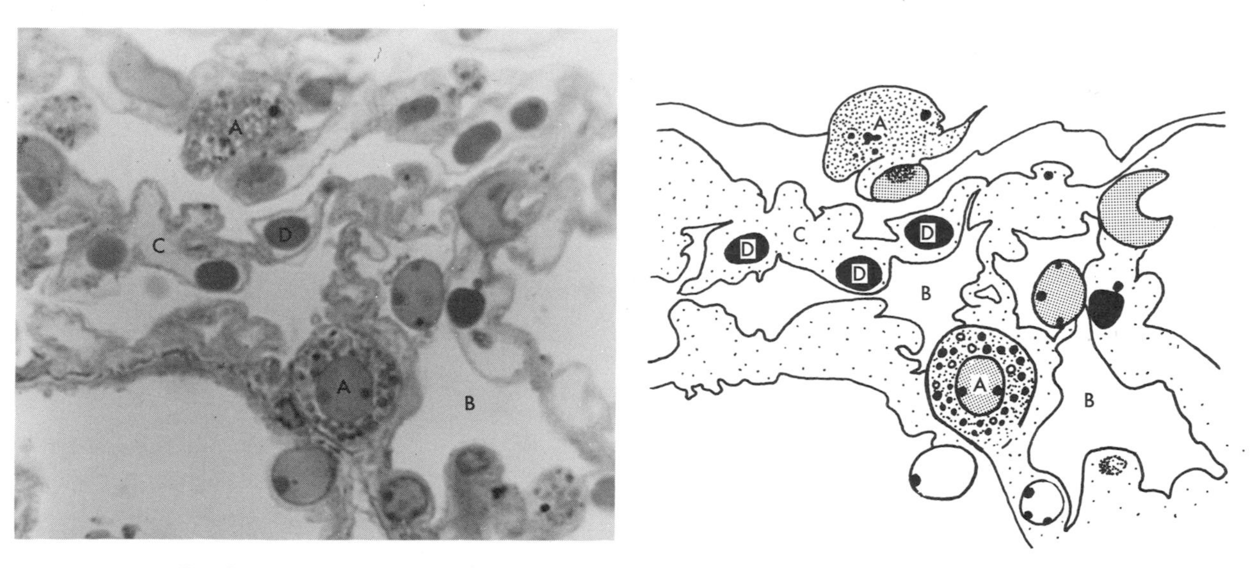

Fig. 3-2. Lung tissue, showing the microscopic structure of an alveolus. Left: Photograph of lung tissue; right: drawing of same picture, illustrating important structures. *A,* Macrophage cell; *B,* lumen of alveolus; *C,* blood capillary; *D,* red blood cell. (Magnification X1000.) (Photograph courtesy Dr. Warren Andrew, Indiana University Medical School, Indianapolis.)

together by connective tissue and resemble a large block of sponge rubber.

Physiology of gas exchange. The atmosphere we breathe is about 21% oxygen, 79% nitrogen, 0.03% carbon dioxide, small amounts of inert gases, and variable amounts of water vapor (also a gas). Under normal conditions the nitrogen of the air is completely inert to all organisms except some bacteria. The percentage of the gases in the atmosphere is extremely constant, regardless of elevation, latitude, or longitude. However, the concentration (number of molecules per unit volume) does vary, particularly with high elevations or great depths.

Carbon dioxide transport. In addition to being a waste product of metabolism, carbon dioxide (CO_2) has several important physiological functions. Its effects are (1) regulation of breathing rate, (2) maintenance of the proper pH (acidity) in the tissues, (3) increased release of oxygen from hemoglobin, and (4) indirect control of blood pressure and heart rate. These effects are to a large degree due to the chemical reactions that CO_2 undergoes. When CO_2 is released into the blood from the tissues, it can be carried as CO_2 dissolved in water but is usually converted to carbonic acid (H_2CO_3). The enzyme *carbonic anhydrase,* abundant in red blood cells, greatly accelerates this reaction as follows:

$$CO_2 + H_2O \xrightarrow{\text{Carbonic anhydrase}} \underset{\textbf{Carbonic acid}}{H_2CO_3}$$

When H_2CO_3 is formed, it in turn ionizes into hydrogen ion (H^+) and bicarbonate ion (HCO_3^-).

$$H_2CO_3 \rightleftharpoons H^+ + HCO_3^-$$

The bicarbonate ion then associates with sodium or potassium ions (to make $NaHCO_3$ or $KHCO_3$), and the H^+ ion is bound to hemoglobin. These reactions are freely reversible, depending on prevailing concentrations of the individual compound, such that when these compounds reach the lungs, CO_2 is released into the atmosphere.

Oxygen transport. Oxygen does not dissolve readily in water, but this problem is solved by the presence of oxygen carriers in the blood. There are many different carriers, but they all have one common characteristic: each contains a metal combined with a protein in such a way that oxygen can be bound reversibly. The most familiar oxygen carrier is hemoglobin, which contains 574 amino acids and has the following general formula: $C_{3032}H_{4816}O_{872}N_{780}S_8Fe_4$. It is composed of a protein (globin) to which four heme groups are attached (Fig. 3-3). In the lungs, oxygen moves across the alveolar membrane and dissolves in the blood. If it were to remain in solution, the blood would soon become saturated with oxygen. Instead, the oxygen loosely binds to hemoglobin in the blood to produce a compound known as oxyhemoglobin. This phenomenon allows the blood to pick up large quantities of oxygen. Hemoglobin combines with oxygen when the free oxygen concentration is high and releases it when the oxygen concentration is low.

The blood leaves the lungs (where the oxygen concentration is high) and moves into the body tissues (where the oxygen concentration is low). In the capillaries of the tissues the oxygen is released from the hemoglobin molecules. Then it diffuses from the capillaries into the body cells, where it is consumed during the last step in the electron transport system of cellular respiration.

Contaminated air. Since the air we breathe is rarely free of contaminating particles, the respiratory tract has developed mechanisms for handling this foreign material. Particles larger than 10 μm in diameter usually are stopped either by the hairs or mucous membranes of the nasal passages. Those particles ranging in diameter from 2 to 10 μm are usually caught by a sheet of mucus lining the trachea, bronchi, and bronchioles (Fig. 3-4). Although these particles can be removed by a forceful blast of air such as a cough or sneeze, they are more often removed by the action of a sheet of mucus secreted by mucous glands lining the upper respiratory tract.

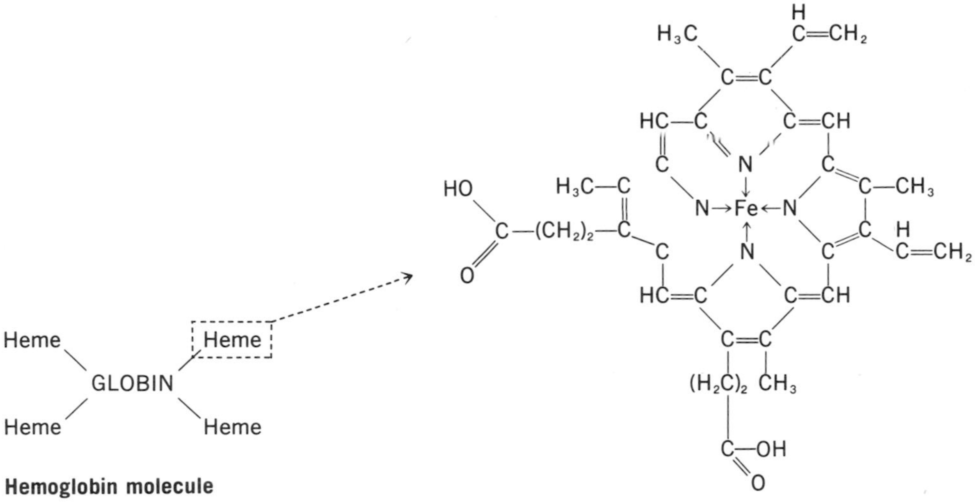

Hemoglobin molecule

Structure of heme group

Fig. 3-3. Structure of hemoglobin. Hemoglobins of vertebrates consist of four iron-containing (heme) groups attached to a larger protein "carrier" portion of the molecule, the globin. Each heme group can carry one oxygen molecule—a total of four for the molecule. The atom of iron (Fe) in the heme group acts as the site of the attachment of an oxygen molecule.

The cells of this tract have hairlike structures called cilia, which move in a wavelike motion, in turn, moving the mucus upward at a rate of about 1 inch per minute. The overall effect of this constant motion is to move the mucus sheet with its trapped particles from the bronchioles, bronchi, and trachea up to the pharynx, where it can be removed by spitting or, more often, by unconscious swallowing.

It is unlikely that any particle larger than 2 μm will make its way into the alveoli. Particles smaller than 0.3 μm are removed by exhaling if they remain suspended in the air; otherwise, they may be taken up by the blood. Therefore, the particles that remain in the alveoli are usually in the 0.3 to 2 μm size range. Some of these particles may remain in the alveoli permanently, as is indicated by the dark-colored lungs of miners or those individuals who have lived in areas with severe air pollution. The alveoli, which have neither cilia nor mucous glands, have other ways of handling these contaminants. There are many cells throughout the body that have an affinity for particulate matter and other cellular debris. These cells, called *macrophages,* are of unknown origin, but some of them appear to be derived from certain types of white blood cells. Because of their mobility and phagocytic activity (ability to ingest solid matter), they act as scavengers that engulf dead cells, bacteria, and foreign matter. The ingested organic material is then destroyed by the action of their intracellular proteolytic enzymes, but the inert foreign matter that resists digestion may remain in the cells indefinitely. Certain of the white blood cells move outside the circulatory system showing active movement and exhibiting phagocytosis in a manner similar to that of the macrophages. Both types of cells can surround foreign particles in the air ducts and be carried up the respiratory tract by ciliary action. They can then be removed by spitting or swallowing.

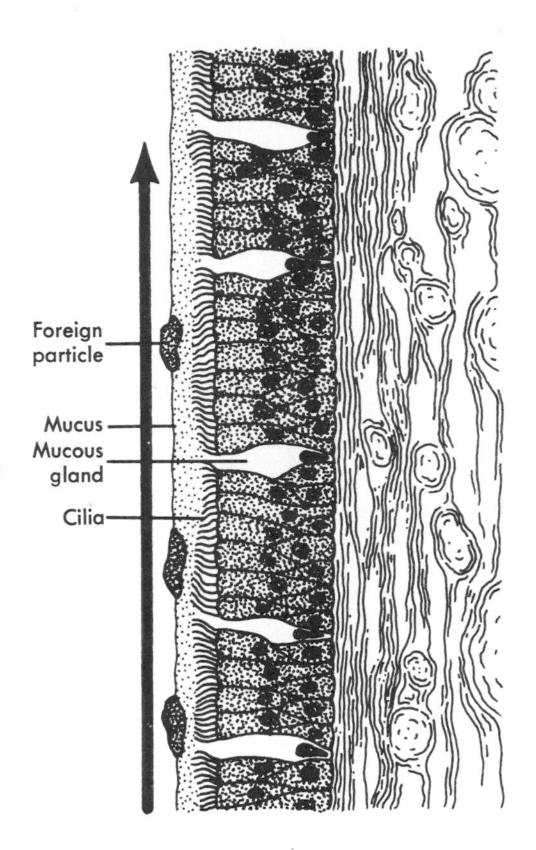

Fig. 3-4. Mucus sheet in the upper respiratory tract. The cells lining the bronchioles, bronchi, and trachea are ciliated. The wavelike motion of the cilia moves a thin sheet of mucus (produced by the mucous glands) upward at a rate of about 1 inch per minute. Foreign particles in the air are trapped in the mucus and removed from the lungs.

CIRCULATION AND THE BLOOD

In addition to supplying a continuous flow of blood to the tissues the circulatory system has other functions. It carries oxygen and carbon dioxide to and from all cells of the body, produces antibodies that combat disease, transports waste products to the kidneys, carries hormones from sites of production to sites of action, carries excess heat to the cooler regions of the body, and aids in repair of damaged tissue. Since the circulatory system serves so many important functions, perhaps ancient peoples were not so far wrong in considering the heart as the center of the soul!

The circulatory system consists of a pumping device and a series of connecting vessels arranged in such a way that the internal fluids are pumped in one direction, passing through the body tissues before returning to the pumping device. The circulatory systems of all vertebrates, including man, are built along a common plan (Fig. 3-5). *Arteries* carry blood from the heart to smaller vessels called *arterioles*. These connect to minute thin-walled vessels, the *capillaries*. After passing through the tissues the capillaries reunite to become slightly larger *venules*, which in turn unite to form *veins* and return the blood to the heart.

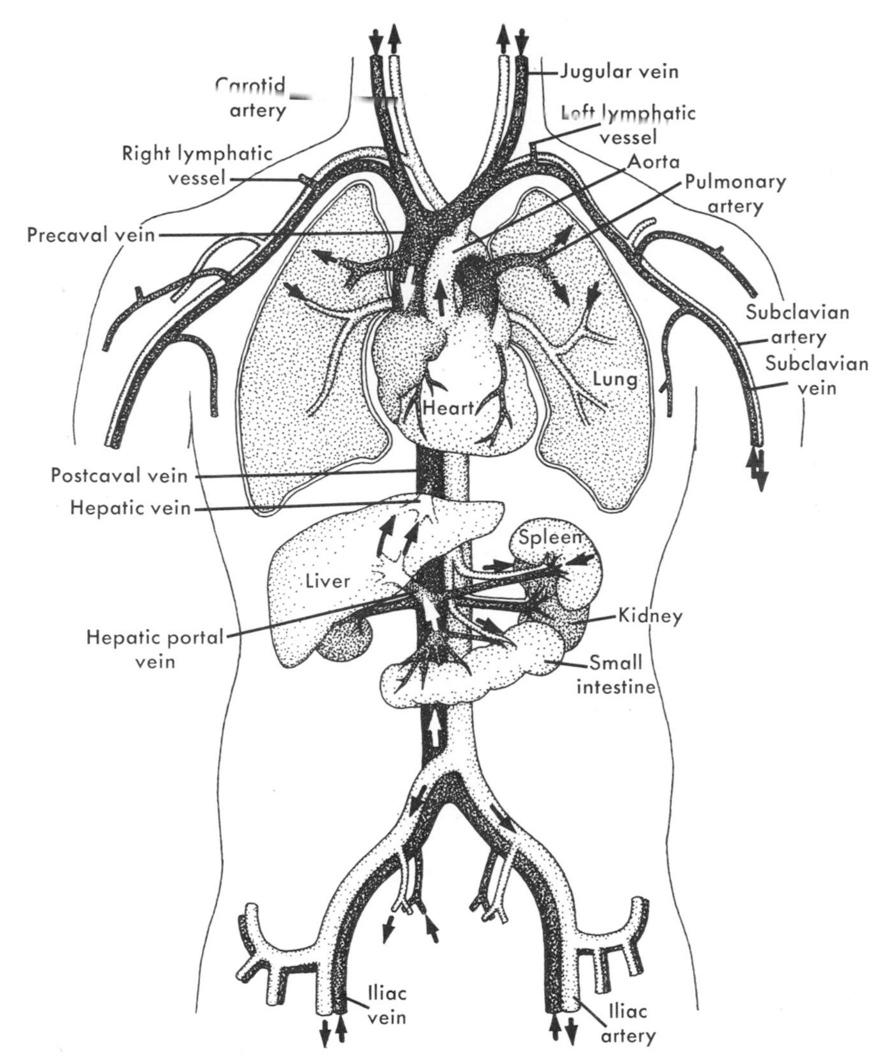

Fig. 3-5. Human circulatory system. This system is representative of all the mammals.

The capillaries are perhaps the most important of the vessels, since it is through their walls that all of the exchange of materials between the blood and tissues occurs. A capillary is a small tube, usually about 10 μm in diameter, composed of a single thickness of cells bound together at their edges with a cellular glue. The capillaries are so numerous that they penetrate to the individual cells of every tissue. It is thought that no cell is more than one cell's thickness away from a capillary. This means that there may be as many as 1.5 million capillaries per cubic inch of tissue.

A small muscular ring (precapillary sphincter) around the entrance, the arterial end, to the capillary controls the flow of blood into the capillary network. Only a small percentage of the total number of capillaries is usually open at any one time. Because the walls of arteries and veins are impermeable to substances in the blood, all exchange of materials must take place across the thin capillary walls.

There is a net outward movement of nutrients, gases, water, and proteins at the arterial ends of capillaries and a net inward movement of wastes, gases, water, and proteins at the venous ends. Overall, there is little difference in the volume of materials that leave the capillary and those that reenter the capillary. This balance is dependent on the blood pressure. If the blood pressure is too high, too much fluid leaves the capillary; and if the pressure is too low, not enough fluid leaves.

Lymphatic system. In addition to the arterial and venous systems the higher vertebrates have a second circulatory system of great importance: the lymphatic system. In mammals the regular circulatory system is a closed, high-pressure system that includes the extensive capillary network. Although the movement of materials between the capillaries and tissues is normally well balanced, there are some materials that do not return to the capillaries. A "drainage system" is required, and the lymphatic system meets this need. Between the cells in most of the tissues there are minute channels where fluid (lymph) collects. These channels converge to form lymph vessels (Fig. 3-6). The vessels unite to form a large duct that drains most of the body and empties into the left subclavian vein, thereby returning the lymph to the general circulation. Lymph from the right side of the head, neck, and chest empty into the right subclavian vein. The lymphatic system also functions in removal of foreign or damaged material from the blood, for example, foreign materials that enter the blood through the lungs. *Lymph nodes,* located at intervals along the larger lymphatic vessels, function in the production of antibodies and as barriers to the spread of bacteria through the body.

Vertebrate blood. The components of vertebrate blood are (1) a straw-colored *plasma* that makes up about 55% of the total blood volume, (2) *erythrocytes,* or red blood cells, containing hemoglobin, (3) several types of *leukocytes,* or white blood cells, and (4) very small *platelets* (Fig. 3-7).

Plasma is about 90% water. The total concentration of salts (NaCl, KCl, $NaHCO_3$, Na_2HPO_4, NaH_2PO_4, etc.) in the plasma is about 0.9%. Three major proteins, *serum albumin, serum globulin,* and *fibrinogen,* make up about 6% to 8% of the plasma. The main contribution of serum albumin is its osmotic effect in causing fluids to reenter the capillaries. The serum globulins (particularly gamma globulin) are important in the development of immunity. Fibrinogen is important in blood clotting. The plasma proteins are important as a source of nitrogen-containing nutrient materials (e.g., amino acids). Plasma also contains a small amount of glucose (blood sugar).

In all vertebrates except the mammals erythrocytes are nucleated and oval shaped. Mammals typically have nonnucleated, biconcave, circular red blood cells (Fig. 3-7). Immature mammalian erythrocytes have a nucleus, but it is lost before the cell enters the bloodstream. In the human fetus the liver, spleen, and lymph nodes all produce red blood cells. Until adolescence, the marrow of the long bones of the arms and legs produce red blood cells, but by the age of 20 years only the marrow of the membranous bones (sternum, ribs, and vertebrae) produce the cells. Mature erythrocytes average 7.5 to 8 μm in diameter and function for about 100 to 110 days. Human blood averages about 5.4 million red blood cells per cubic millimeter in men and 4.8 million per cubic millimeter in women. The most important component in the erythrocytes is hemoglobin. This respiratory pigment enables the blood to carry 20 ml of oxygen for each 100 ml of blood. Were the hemoglobin free in the blood and not contained within the red cells, the blood would be much more viscous and difficult to circulate.

There are several different kinds of leukocytes, or

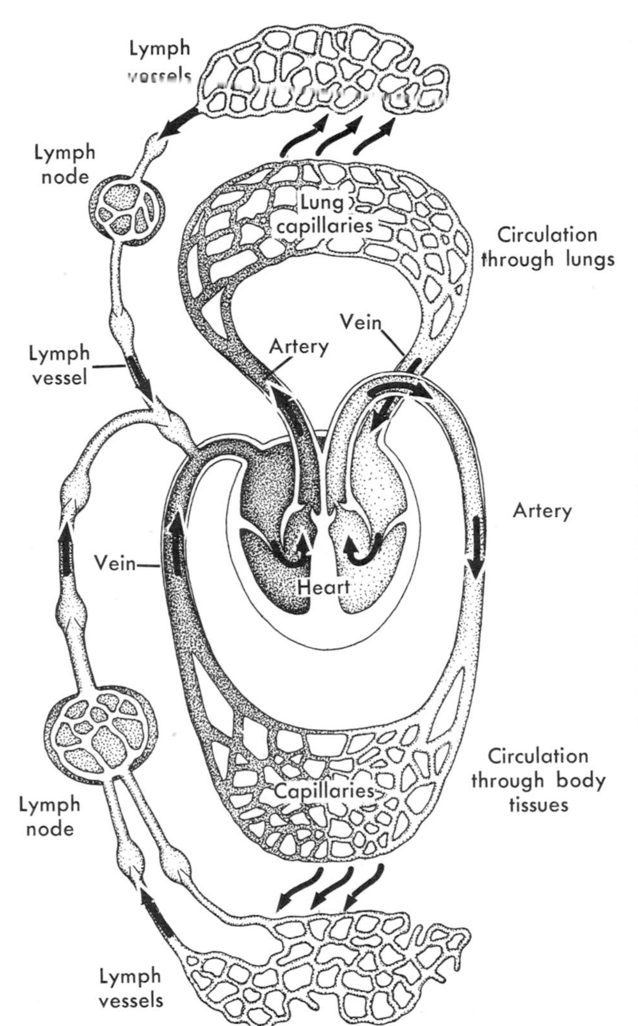

Fig. 3-6. This schematic diagram shows the relationship between the arterial and venous systems and the lymphatic system. Blood is pumped by the heart through a network of capillaries in the lung. The oxygenated blood returns to the heart, where it is then pumped out to supply the various parts of the body with oxygen and nutrients. After passing through the tissues the blood returns to the heart to again be pumped to the lungs. However, in both the lung circulation and body tissue circulation some of the fluid leaves the blood capillaries and enters spaces within the tissues. This fluid is picked up by the lymph capillaries and returned to the bloodstream. Lymph nodes are important in the removal of foreign particles and in the production of antibodies.

white blood cells, but the function of most of them is the destruction or inactivation of foreign particles. Although there are large numbers of them circulating in the blood, by far the greater number is present in the tissues. Under normal conditions there are about 5000 to 10,000 white cells per cubic millimeter of blood. The two major categories of leukocytes are those that have coarse granules in their cytoplasm *(granulocytes)* and those that have smooth-looking cytoplasm *(agranulocytes).* Granulocytes can be further divided into *neutrophils, eosinophils,* and *basophils.* These are separated by the way in which they stain in blood preparations. Specific changes in numbers of the various types do occur in many diseases, but the causal relationship is still unclear. The two important types of agranulocytes are the *lymphocytes* and *monocytes.* It is thought that lymphocytes may be transformed into monocytes under certain conditions. Both lymphocytes and monocytes are capable of releasing antibodies, but only the monocytes possess phagocytic powers, being able to engulf and destroy bacteria or other foreign bodies. All the white cells are capable of at least some ameboid movement. This enables the cells to penetrate the capillary wall and go into the tissue spaces. Granulocytes are produced in the bone marrow, and lymphocytes and monocytes are formed in the lymph nodes, spleen, thymus gland, and sinuses of many organs. Average percentages of the leukocytes follow: neutrophils, 55%; lymphocytes, 30%; monocytes, 7%; basophils and eosinophils, 8%.

Platelets, sometimes called thrombocytes, are non-nucleated, irregular-shaped oval bodies about 3 μm in diameter. When a blood vessel is damaged, the platelets disintegrate and release a substance that is important in blood clotting. There are usually about 500,000 platelets per cubic millimeter of blood, and they are produced by the bone marrow.

Antigens and antibodies. Several substances in the blood, especially the plasma globulins, react against infective organisms in the body. The defense against these foreign organisms is a chemical one and involves the formation of *antibodies.* For a long time it has been known that after a person became infected with a particular disease he never contracted the disease again; that is, he had developed an *immunity* to the disease. Cowpox and smallpox are caused by viruses so closely related that immunity to one gives immunity to the other. Near the close of the eighteenth century Edward Jenner successfully immunized peo-

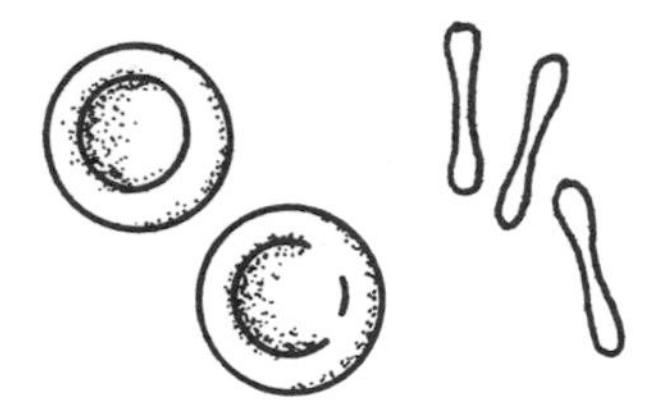

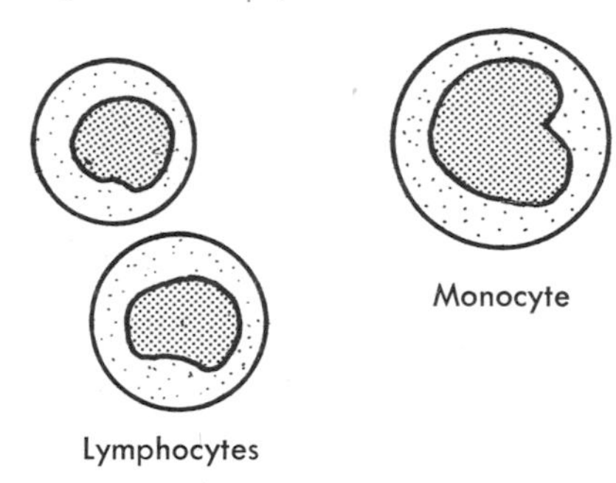

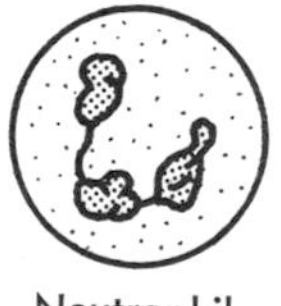

Fig. 3-7. Formed components of vertebrate blood. Notice that the human red blood cell has no nucleus. It is lost as the cell matures. Platelets occur in all mammals and function in blood clotting.

ple against smallpox by vaccinating them with fluid taken from cowpox vesicles, which contained the active virus. This was the discovery on which all inoculations are based. Today the production of antibodies is stimulated against a particular disease by injecting weakened foreign proteins, killed pathogens, or strains of reduced virulence bred specifically for this purpose.

The initial step in the immunity reaction is introduction of foreign proteins (bacteria, fungi, bites, stings, etc.) into an organism. These foreign proteins are called *antigens.* The antigens cause the *gamma globulins* of the plasma to produce *antibodies* against the specific antigen. For example, antigen A will stimulate the formation of antibody A; whereas antigen B will have no effect on antibody A. The antibodies are produced when the antigen enters the lymphatic tissue (lymph nodes, spleen, and thymus). Antibodies may remain active for a variable period of time. Thus after the organism has been invaded by an antigen, specific antibodies are capable of destroying the antigen on subsequent invasions, depending on the type of antibody. Immunity to the common cold virus lasts only a few weeks, whereas immunity to measles lasts a lifetime.

DIGESTION

Living organisms can be loosely classified as *autotrophic* or *heterotrophic.* The autotrophic organisms (green plants and some bacteria) can synthesize complex materials from inorganic mineral sources. Animals are heterotrophic organisms; that is, they depend on organic materials as a food source and have relatively limited synthesizing ability. It appears that the nutritional requirements are similar throughout the Animal Kingdom. Metabolic pathways are almost interchangeable, and the nutrients, on the whole, are also interchangeable. Whether the food consists of plant or animal matter, it usually contains proteins as well as fats and carbohydrates. Some of these compounds, particularly the carbohydrates and fats, are used to meet energy requirements. The components of protein, amino acids, are required to build and repair the organism's tissue. The process of preparing these organic molecules for use in cellular metabolism is called digestion.

Sources of energy. Carbohydrate is normally the major source of energy for the cell—about 50% to 75% of the total calories used are derived from this source. Carbohydrates (composed of carbon, hydrogen, and oxygen) include sugars and starches, and the usual sources of these materials are potatoes, rice, cane sugar, or any "starchy" food. Carbohydrates range in complexity from the very simple to molecules with molecular weights of several millions. The complex molecules are generally reduced to simple sugars (such as glucose) by digestive enzymes before entering the metabolic pathways of the cell.

Conversion of excess carbohydrates into fat is also a common process. Fats can be utilized as an energy source just as are the carbohydrates. Fats are stored when food is abundant, but when required, fats are broken down and thereby release energy for use in metabolic processes. Many animals (e.g., the migrating salmon) can live for months on reserves of accumulated fat. Since fat provides over twice as much energy per gram as do other energy sources, it is apparent why animals store their excess energy as fat rather than as carbohydrate or protein. Common sources of fat are vegetable oils, butter, animal fat, and cream. Although the primary importance of protein is related to its role in building new body tissue, it can also be used as an energy source. If proteins are to be utilized for energy, they are first broken down by digestive enzymes to their component parts, the amino acids. These amino acids are then deaminated (removal of the amino group, $-NH_2$) and subsequently oxidized in the citric acid cycle reactions. Common sources of protein are meat, milk, eggs, beans, and gelatin.

Divisions of the digestive system. Posterior to the mouth is the pharynx (Fig. 3-8), a region shared jointly by the respiratory and digestive systems. The

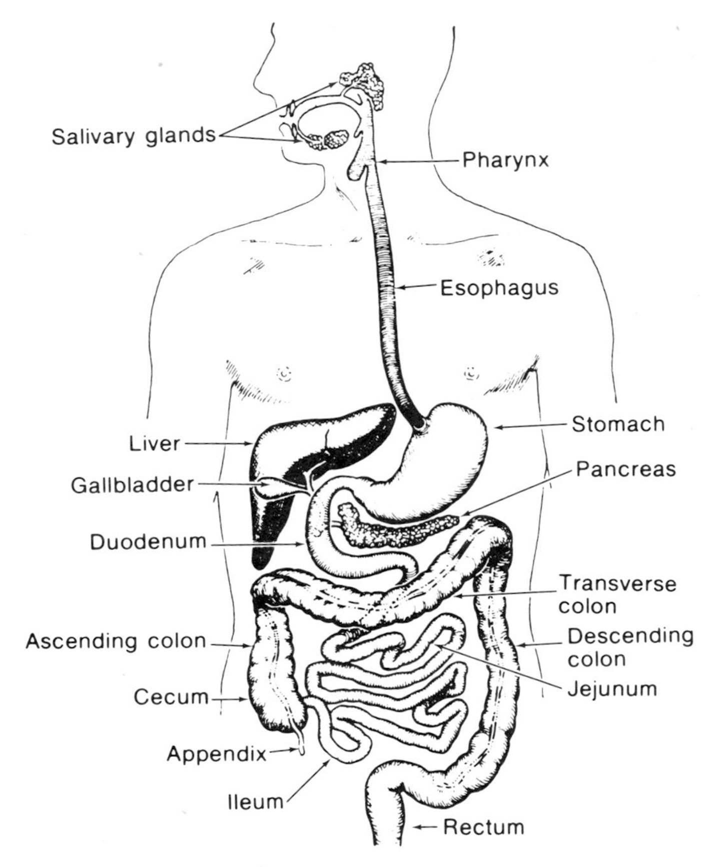

Fig. 3-8. Human digestive system. (Modified from Schottelius, B. A., and Schottelius, D. D.: Textbook of physiology, ed. 17, St. Louis, 1973, The C. V. Mosby Co.)

pharynx is joined to the stomach by the esophagus, whose only function is to move food to the stomach. Once in the stomach, the lumps of food are transformed by mechanical action into a soft homogenous mass called chyme. Glands in the stomach wall secrete digestive enzymes, hydrochloric acid, and mucus. The stomach leads into the small intestine through a muscular valve. The duodenum is the first and widest part of the 20-foot-long (in man) small intestine, and it is the site of the final digestive processes. Absorption of digested food into the bloodstream occurs in the remaining two parts of the small intestine, the jejunum and the ileum. The greatest specialization in the small intestine is the development of countless microscopic projections called *villi,* covering the entire inner surface (Fig. 3-9). The blood and lymph vessels that run through the villi function in absorption. Compared to the small intestine, the large intestine is a large, thin-walled sac, much wider in diameter and more distensible. Three divisions, totaling about 5 feet in man, can be distinguished. The first part, or cecum, is a pouchlike structure from

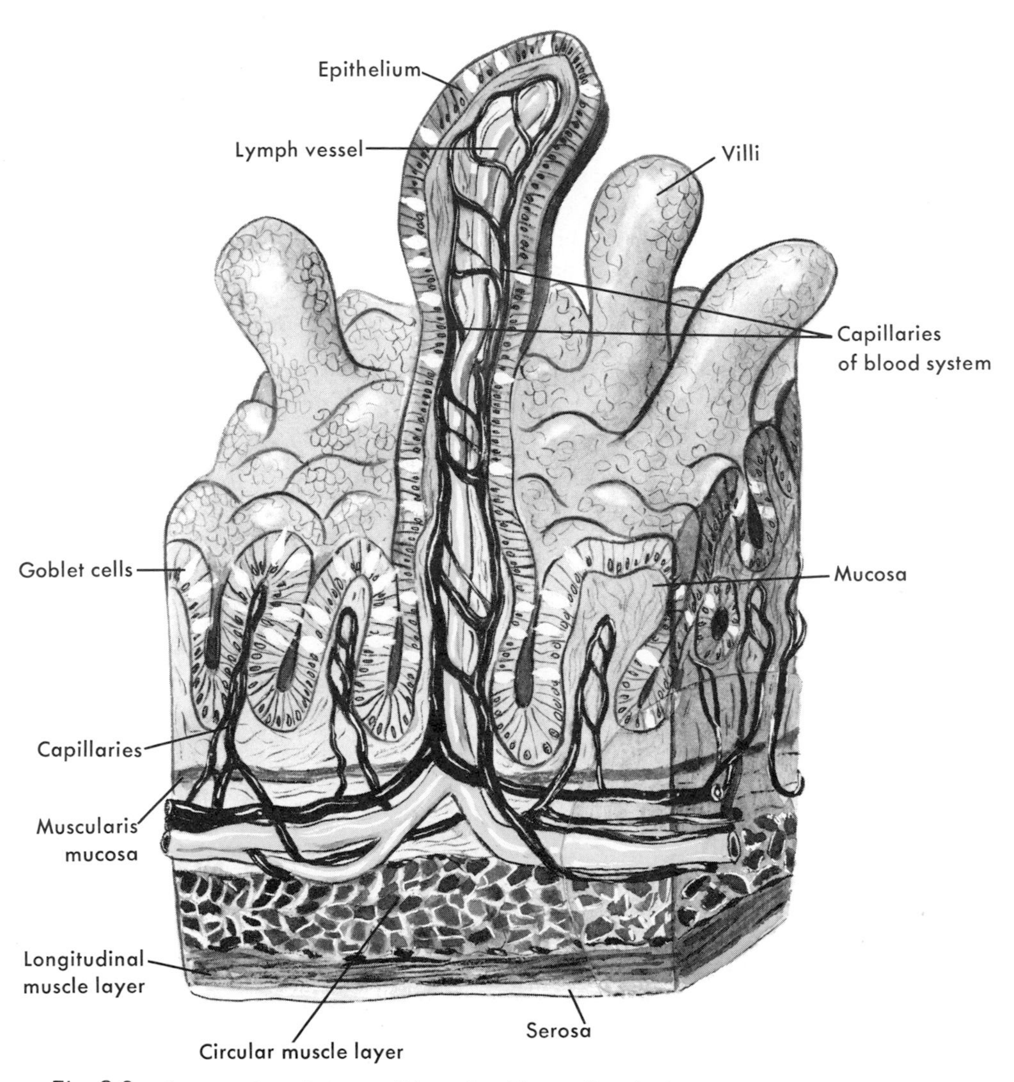

Fig. 3-9. Cross section of the small intestine. The small projections, or villi, from the surface are responsible for absorption of food materials. As digested food is transported out of the gut cavity, it is picked up by the vessels of either the blood or lymphatic system. (Modified from Schottelius, B. A., and Schottelius, D. D.: Textbook of physiology, ed. 17, St. Louis, 1973, The C. V. Mosby Co.)

which hangs the appendix. The second part is the colon, which joins the last part of the large intestine, the rectum, which opens to the exterior through the anus.

Chemical breakdown of food. Digestive enzymes are secreted in several parts of the digestive system, and each type of enzyme secreted reacts with a specific type of food molecule. The overall function of digestion is to break down large complex food molecules (carbohydrates, lipids, and proteins) into their simple components, which can then be absorbed. The intestinal wall is almost impermeable to complex carbohydrates, but on complete digestion three kinds of simple sugars result: glucose, fructose, and galactose. Of these three, glucose is by far the most abundant (about 80% of the total) and is absorbed by the intestine; the other two each constitute about 10% of the total. Galactose and fructose are absorbed into the circulatory system, carried to the liver, and there transformed into glucose. This glucose may then be utilized in the cells for the production of energy or for the rebuilding of complex carbohydrates.

Lipids (for example, fats) are emulsified by bile, which is secreted into the intestine from the liver before being acted on by the lipid-digesting enzymes from the pancreas. On complete degradation, fat molecules produce fatty acids and glycerol, which are absorbed through the gut.

Proteins are formed by the combination of approximately twenty different kinds of amino acids. Most proteins contain several thousand of these amino acids combined in various sequences to produce distinctive proteins. When acted on by the appropriate enzymes from the stomach, pancreas, and intestine, all of these protein molecules are broken down to their component amino acids, which can then be absorbed into the bloodstream.

Absorption. When digestion is complete, all carbohydrates are reduced to monosaccharides (simple sugars), all fats to fatty acids and glycerol, and all proteins to amino acids. Actually, complete digestion rarely occurs, and there is some absorption of food materials that are intact or only partially broken down. About 90% of the total absorption occurs in the small intestine, but some absorption of certain materials does occur in other regions. The stomach can absorb small amounts of glucose, alcohol, salts, and water. Once the material passes through the intestinal wall, it enters the circulatory system or the lymphatic system.

Materials can be absorbed by *diffusion, active transport,* or *phagocytosis.* Diffusion is a passive process and requires no energy. The cell does not have to supply energy for the process. Active transport requires that the cell chemically react with the substance and move it "uphill," so to speak. Active transport requires an energy expenditure. Phagocytosis involves a cell engulfing a solid particle, and this also requires the cell to use energy.

Some monosaccharides (e.g., fructose) are absorbed almost exclusively by diffusion, whereas others (glucose, galactose) are actively transported across the intestinal wall. Although the primary method of fat absorption is probably diffusion, electron microscopy has suggested that some fat may be taken in as micro-droplets by phagocytosis. There is some absorption of unbroken fat molecules into lymph vessels in the intestinal wall. Amino acids can diffuse passively through the wall of the intestine; indeed it may be the only method of absorption for some amino acids. However, most amino acids are actively transported out of the intestine. In addition to food molecules, other materials must also be taken in. Water is absorbed by simple diffusion and requires no biochemical work. Sodium is actively transported out of the intestine, and it is believed that potassium, calcium, magnesium, phosphates, and iron are actively transported also.

KIDNEY FUNCTION

One of the major problems of animal life is to maintain inside the organism just the proper amount of water and dissolved materials–not too little, not too much. The environmental range of water and salts from fresh water through the sea to salt lakes, from humid swamps to dry deserts—is far greater than the tolerated range of body fluid concentrations. To meet the problems posed by these environments, efficient regulatory structures are essential. These same regulatory devices, the kidneys, also function in excretion.

The functional unit of the kidney is the *nephron,* of which there are about 1 million in each human kidney. The nephrons empty into collecting ducts, which in turn empty into a central cavity in the kidney called the pelvis. From the pelvis a duct (the ureter) carries the urine from the kidneys to the bladder, from which the urine passes to the outside through the urethra (Fig. 3-10).

Each nephron consists of five parts in the following sequence: (1) Bowman's capsule, (2) proximal convoluted tubule, (3) loop of Henle, (4) distal convoluted tubule, and (5) collecting duct (Fig. 3-11). These structures are arranged in such a manner that the Bowman's capsule and the proximal and distal tubules are near the outer edge of the kidney, and the loop of Henle is hairpin shaped with the bend at the inner portion of the kidney. Each kidney is supplied with a renal artery that branches off the aortic artery. Inside the kidney the artery branches into arterioles, and each arteriole forms a network of capillaries at a nephron. Each capillary network is called a glomerulus and is located inside a cuplike Bowman's capsule. The capillaries of the glomerulus reunite into an arteriole, and the blood flows away from the glomerulus and again branches into a second capillary network around the proximal and distal convoluted tubules as well as the loop of Henle. These capillaries then unite to form the renal vein, which carries the blood out of the kidney back into the venous circulation.

Waste products result from the breakdown of carbohydrates, fats, and proteins. Complete oxidation of carbohydrate and fat occurs in the cells, leaving only carbon dioxide and water, which can be discharged through the lungs in the expired air. The metabolism of protein, however, creates an additional problem. The amino acids, of which protein is composed, contain amino groups ($-NH_2$), which must be removed before the remaining parts of the amino acids can be used in the energy-generating processes of the cells. Unlike carbon dioxide and water, these nitrogen-containing amino groups cannot be discharged through the lungs. Instead, they are excreted through the kidneys after being converted to either ammonia, urea, or uric acid.

Although most animals produce a small quantity of ammonia, it is an important excretory product only in aquatic invertebrates and fishes. Because it is highly toxic, it must be excreted as rapidly as it is formed. Urea is the major nitrogenous waste product in man (and most other mammals). It is less toxic than ammonia, and moderate amounts of it can be stored in the body for short periods of time. The third important nitrogenous waste product is uric acid. With a few minor exceptions it is the only nitrogenous excretory product that can be excreted in solid form, with almost no loss of water. All of the successful arid-living animals (insects, birds, reptiles) excrete uric acid. Although animals may excrete most of their nitrogenous waste in only one of the three forms, they usually produce small quantities of the other two as well. For instance, humans produce large quantities of urea, but we also produce small amounts of ammonia and uric acid. Differences in the relative percentages of these three waste products may be indicative of certain diseases.

Function of the nephron. The functions of the excretory system are regulation of water and salt and ridding the body of nitrogenous waste. In terrestrial mammals this means that the kidney must remove urea from the circulatory system and deposit it outside the body and, at the same time, try to conserve water and salt. Urine formation in the nephron involves three major processes: filtration, reabsorption, and tubular secretion. Blood pressure in

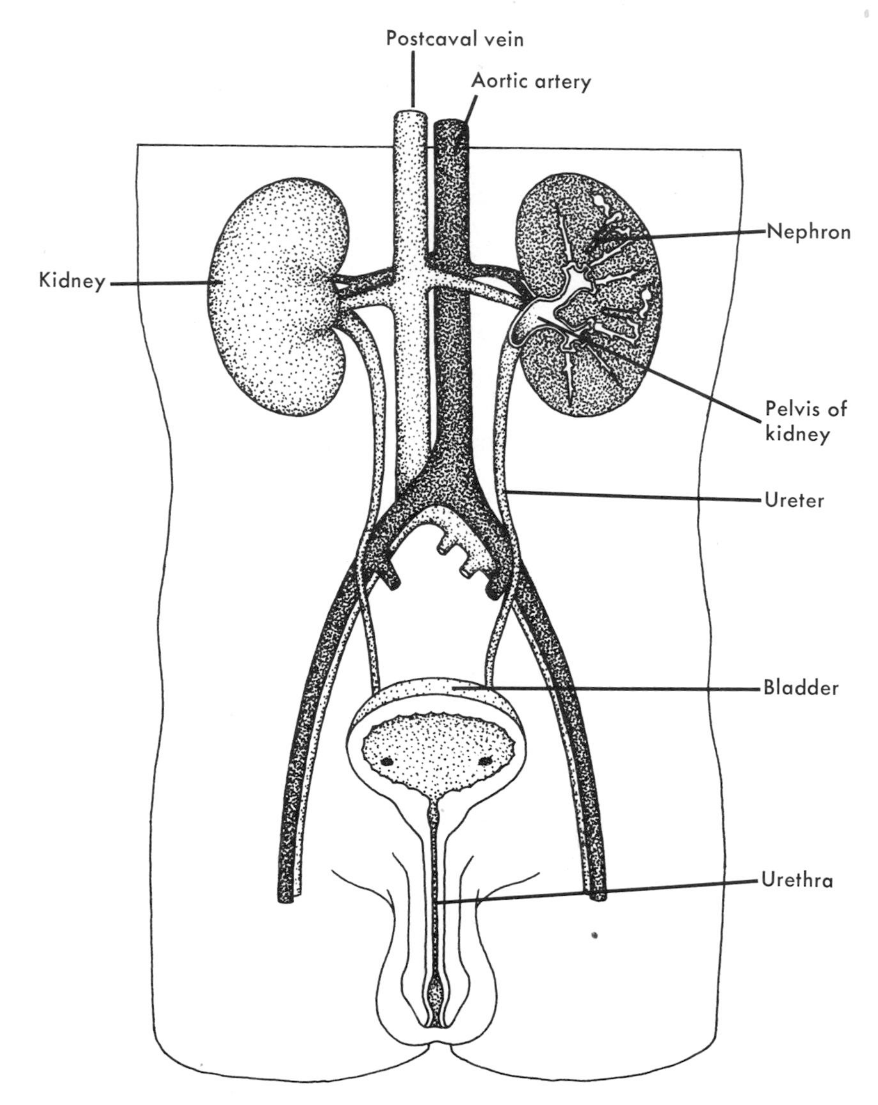

Fig. 3-10. Diagram of the human excretory system, including the principal anatomical features of the kidney. Left, the pattern of circulation within the kidney; right, the arrangement of nephrons within the kidney.

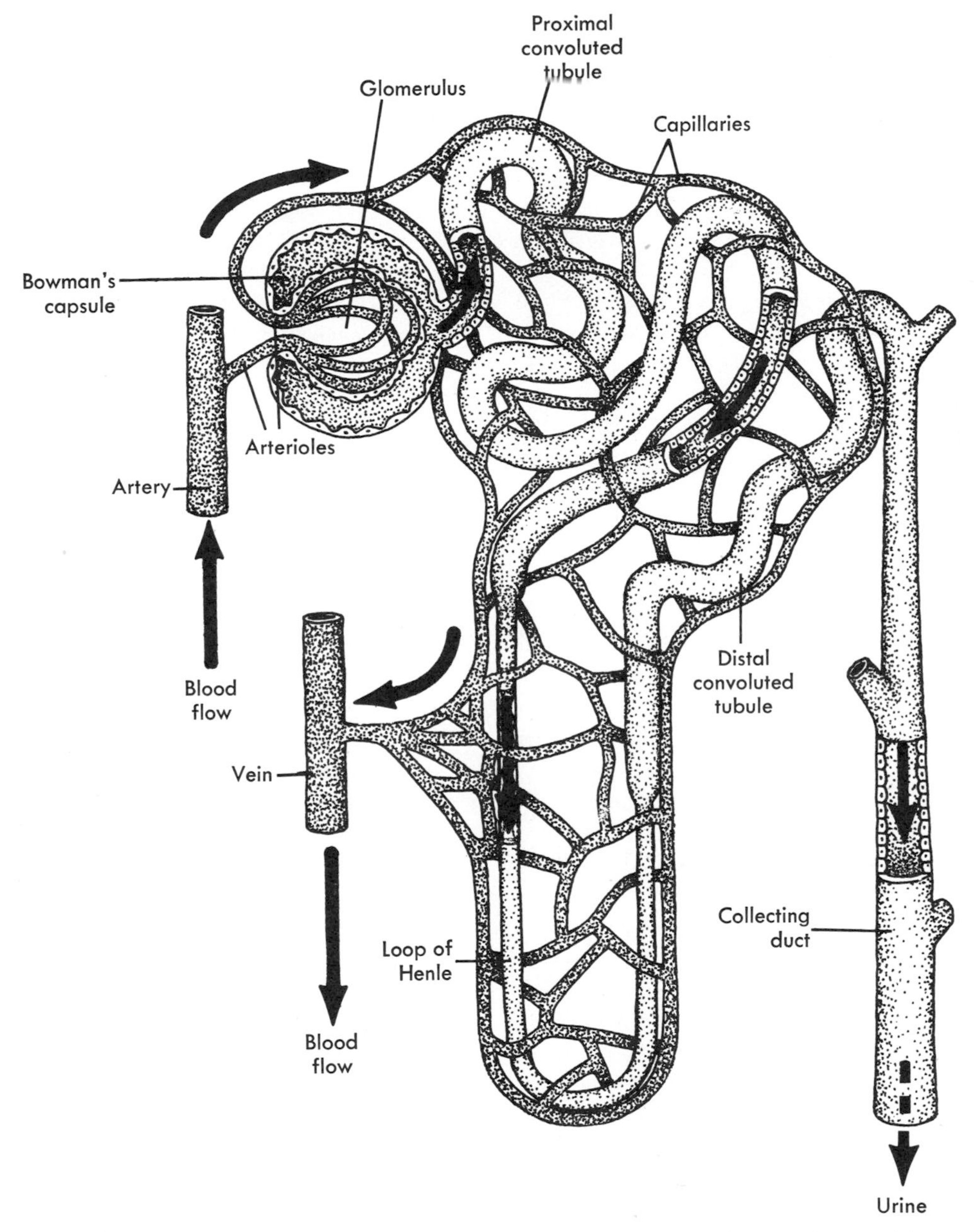

Fig. 3-11. Nephron of the vertebrate kidney, showing the structure of the nephron and its relationship to the circulatory system.

the glomerulus forces a plasma-like fluid through the capillary wall and into Bowman's capsule. The large protein molecules and the red blood cells cannot pass through the capillary wall, thus the wall functions as a filter that lets the small molecules out of the capillary and keeps the large ones in. Dissolved salts and small organic molecules, such as urea and glucose, pass readily through the capillary wall and into the filtrate, but plasma proteins of a molecular weight of 60,000 or more are held back. This process is called filtration and is very similar to the passage of fluid out of the capillaries in other parts of the body.

Each day about 170 liters (45 gallons) of fluid pass into the Bowman's capsules and are processed by the human kidneys. If this fluid were expelled from the body without modification, many valuable substances would be lost (not to mention the mechanical difficulties involved in handling this much liquid). Obviously, most of this must be reabsorbed as it continues its passage through the nephron, leaving only about 1.5 liters of fluid to be excreted daily as urine. The remainder of the water and dissolved substances passes into the capillary network surrounding the tubules and is returned to the blood. In the proximal tubule about 80% of the water is reabsorbed osmotically; and glucose, amino acids, potassium, etc., are reabsorbed by active transport. In the loop of Henle massive amounts of sodium are actively transported out of the nephron and back into the blood. Other substances are removed from or added to the filtrate in various regions of the nephron. As the fluid passes through the collecting ducts, more water is removed from the tubule and returned to the circulatory system. Notice that the only material that has not been reabsorbed is urea, which may comprise about 6% of the urine in man. Small amounts of sodium, potassium, chloride, a little uric acid and ammonia, and a variety of other materials are also present. In addition, some materials (creatinine, potassium, hydrogen ions) are actively transported into the tubule. This tubular secretion is the reverse of absorption.

NERVOUS SYSTEM

The ability to respond to stimuli in the environment is a fundamental characteristic of all living cells, but all cells are not equally sensitive. Certain cells and accessory structures specialize in detecting particular kinds of stimuli. These are *receptors.* The information signals generated by these receptors are then transmitted by *conductors* to special organs of response, particularly muscles and glands. These organs are called *effectors* because they produce a characteristic effect in response to a given stimulus. These receptors, conductors, and effectors are the basic components of the nervous system responsible for coordinating an organism's behavior.

The functional conducting cell in the nervous system is the *neuron* (Fig. 3-12). In large animals a single neuron may be several feet long, but its diameter is no more than a few thousandths of a millimeter. Despite their differences in size and shape, all neurons possess certain features in common. Any given neuron has a cell body containing the nucleus, which controls metabolism in the entire neuron, including the threadlike extensions (called processes). Two types of processes extend from the cell body: (1) *dendrites,* which are often short and which conduct a signal toward the cell body, and (2) an *axon,* which is generally long and which conducts the signal away from the cell body. Each neuron normally has only one axon, but it may have several dendrites. Although several subtle features are used to distinguish axons from dendrites, the important functional difference between them is that dendrites can receive a stimulus from other cells but axons cannot, and axons can stimulate other cells but dendrites cannot.

In the nerves of vertebrates two major types of axons are found: *myelinated* and *nonmyelinated* axons. With the exception of the neurons of the brain and spinal cord, both of these types of axons (and some dendrites) are associated with Schwann cells (Fig. 3-13). In nonmyelinated axons the axon oc-

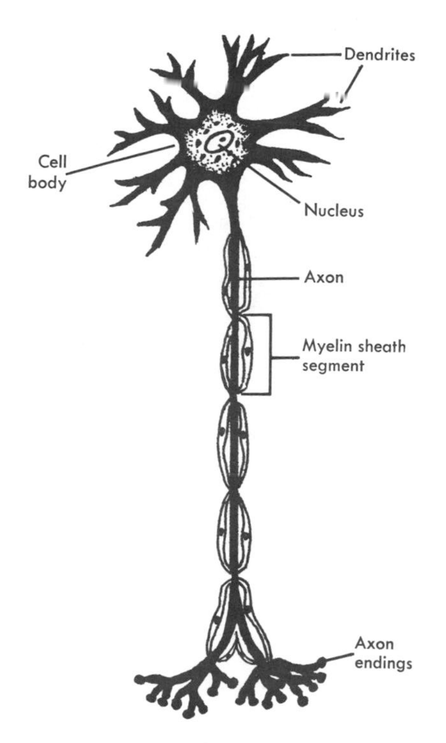

Fig. 3-12. A neuron (diagrammatic). Typically, a neuron consists of one or more dendrites, a cell body that contains a nucleus, and an elongated axon. Although there is great variation in the types of neurons in the different regions of the body, they are all based on this general pattern.

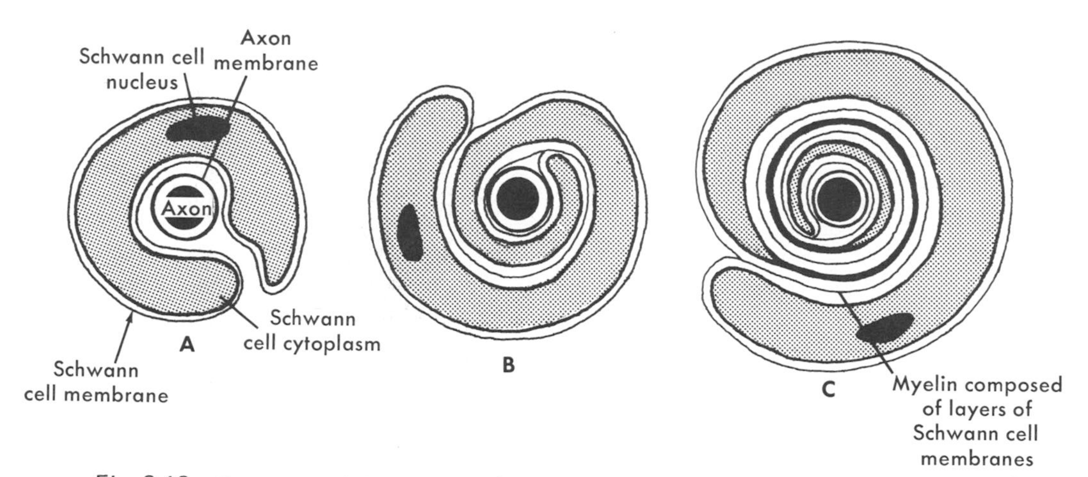

Fig. 3-13. Structure and development of a myelinated axon. Initially, **A**, the Schwann cells merely surround the axon in a simple configuration. Later, **B**, the Schwann cell begins to wrap around the axon, producing, **C**, a multilayered, spiral-like arrangement of its cell membrane.

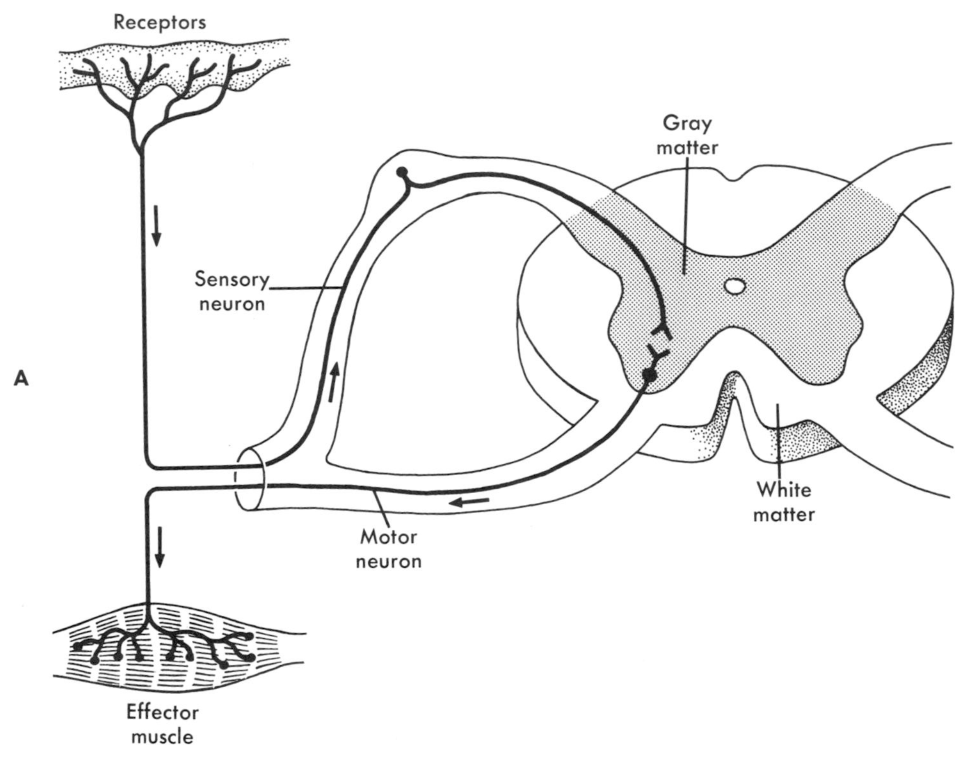

Fig. 3-14. Reflex arc. A, Impulses from the sensory receptors are carried along the sensory neuron to the spinal cord (shown in cross section). The sensory neuron enters the dorsal half of the spinal cord and synapses with a motor neuron in the gray matter of the cord. The motor neuron leaves the cord ventrally and carries the impulse to the effector organ. This is the least complex type of reflex arc. Most are much more complex than this. B, Cross-sectional diagram showing some of the common connections between the sensory and motor neurons. The sensory neuron carries the nerve impulse inward toward the brain or cord. Association neurons can connect directly with motor neurons on the same side of the cord, or they can make multiple connections, crossing over to the opposite side. By means of ascending fibers in the white matter, impulses from the sensory neurons may travel up the cord and finally reach the brain and evoke a sensation. Impulses from the brain are carried down the descending tracts in the white matter of the cord to the appropriate motor neuron and on to the effector organ.

cupies a pocket formed by an indentation in the Schwann cells. In the myelinated axons the Schwann cells wrap repeatedly around the axon, surrounding it with many layers of their cell membranes. This multilayered system formed by the Schwann cells is the *myelin sheath.* The nucleus and cytoplasm of the Schwann cells remain on the outer edge of this multilayered sheath. The material of which the myelin is made is a good electrical insulator. Although the Schwann cells are present in the nonmyelinated axons, they do not produce the myelin sheath. The myelin sheath covering a long axon is not continuous but is composed of segments, each contributed by one Schwann cell.

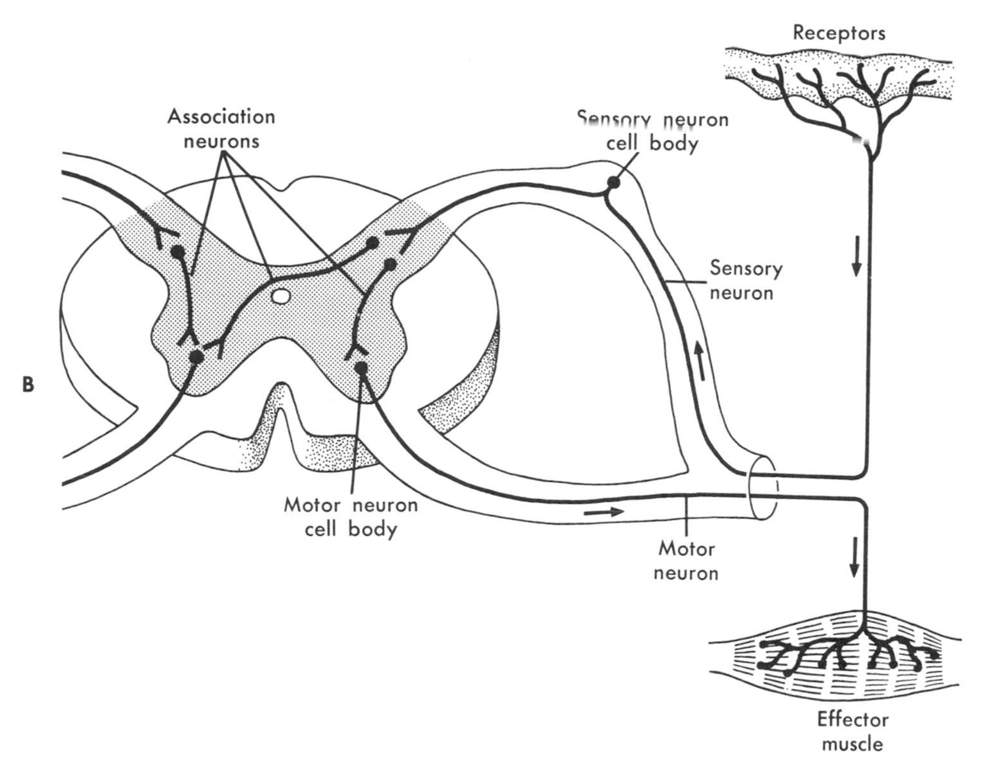

Fig. 3-14, cont'd. For legend see opposite page.

According to their function, the types of nerve cells found in the nervous system include a *sensory neuron,* which receives the stimuli, and a *motor neuron,* which effects a response to stimuli. A *reflex arc* (Fig. 3-14, *A*) in the vertebrate spinal cord demonstrates how a simple nerve pathway functions. The reflex arc transmits an impulse from a sensory receptor to an effector organ. The impulse travels from the receptor to the spinal cord via a sensory neuron. There the impulse is transmitted to a motor neuron. This neuron then carries the impulse to a muscle or gland, where the appropriate response is made. Notice that the brain was not involved in the action. For example, when a person puts his hand on a hot stove, he does not think, "My hand is too hot. I must remove it from the stove." Instead, reflex action involving only receptors (temperature or pain), sensory neurons, motor neurons, and muscles cause him to remove his hand.

How does the person's brain find out that his hand is too hot? Most nerve pathways are more complicated than the one just described. They involve an *association neuron* between the sensory and motor neurons (Fig. 3-14, *B*). It is the association neurons that allow an impulse to be passed to a variety of different effectors, increasing the number of possible responses to a given stimulus. Some branches of the association neurons connect to nerve fibers running to or from the brain. In this way the brain can override a reflex. In the example given, a person

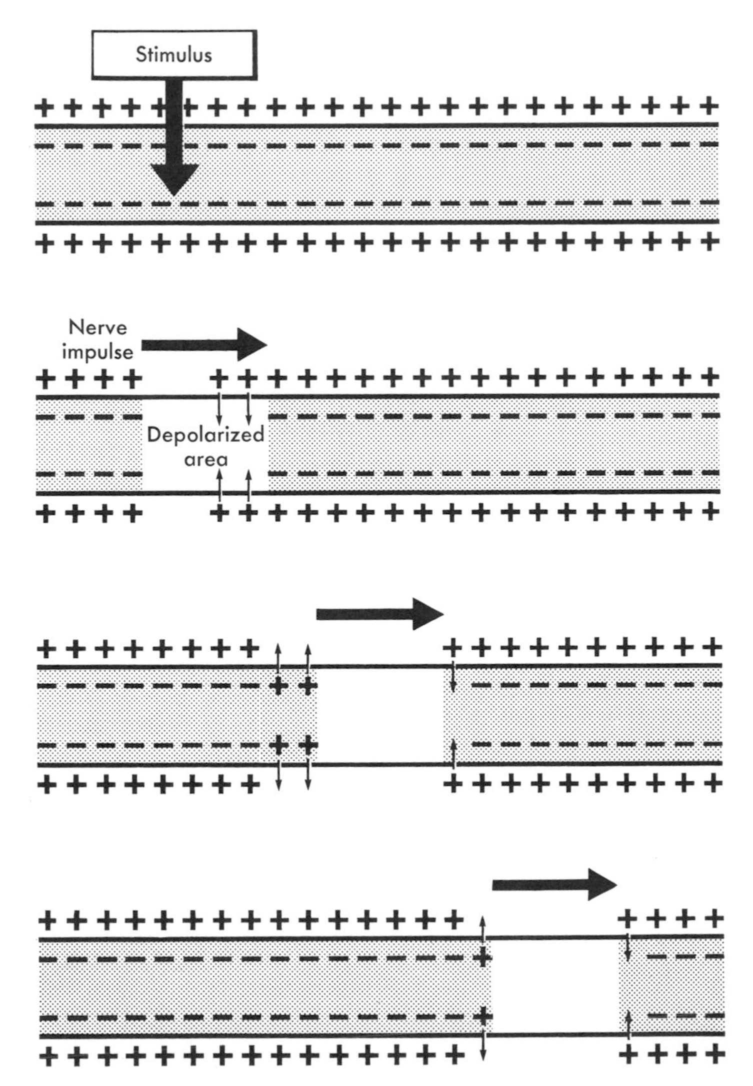

Fig. 3-15. As a result of active transport, sodium ions are concentrated outside the cell membrane of the nerve cell, creating a positive charge on the outside of the membrane while the inside of the cell shows a negative charge. On stimulation the active transport stops, and sodium rushes into the neuron and some potassium moves out of the neuron. This depolarizes the nerve cell, and the depolarization proceeds down the axon. Almost immediately the sodium pump again begins transporting Na^{+} ions across the membrane, and potassium also moves back into the neuron, thus reestablishing the resting potential.

could voluntarily keep his hand on the stove even though it was painful. Obviously, nerve pathways can become exceedingly complex. In vertebrates there are connections between the various spinal nerves, between left and right sides of the body, and numerous connections with the brain. In spite of this complexity, all nerve pathways are built on the pattern represented by the reflex arc.

Nervous activity is associated with a number of physical and chemical nerve fiber changes, none of which is directly visible. When these changes attain a given value, a wave of excitation is transmitted along the fiber. This disturbance is called a *nerve impulse.*

The neuron cell membrane plays a central role in the nerve impulse by separating ions of opposite charge in such a manner than an electrical potential is established between the inside and the outside of the cell (Fig. 3-15). Several factors, including unequal rates of diffusion and active transport, contribute to this ionic imbalance. Sodium and chloride ions are more concentrated outside the cell than inside, and potassium is more concentrated inside. The nerve cell membrane moves sodium out of the cell by active transport. This makes the internal side of the membrane about 70 millivolts negative relative to the outside of the membrane. This polarity is known as the *resting potential.* On stimulation the active transport stops, and there is a rapid influx of positively charged sodium ions to the inside of the fiber. This influx (together with the outward diffusion of potassium ions) temporarily reverses the polarity, giving rise to an *action potential.* Initially, only the point directly stimulated is changed ("depolarized"), but the effect is analogous to a row of tenpins: when one is hit, it falls and knocks the others down in order. Similarly, the whole neuron becomes depolarized as the impulse moves down the axon. Sodium continues to flow into the neuron until the inside becomes positively charged. At this point sodium ions are again actively transported out of the neuron, and potassium moves back into the neuron, reestablishing the resting potential. The nerve fiber is then ready to transmit another impulse. This whole cycle from stimulus to repolarization is accomplished in about 2 milliseconds (0.002 second).

Synaptic transmission. Neurons are able to con-

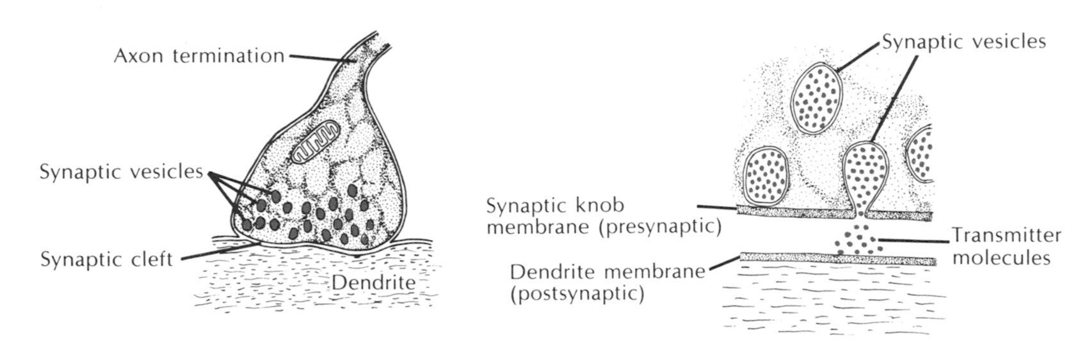

Fig. 3-16. At a synapse the axon membrane (presynaptic membrane) is associated with the dendritic membrane (postsynaptic membrane) of another neuron. When an impulse arrives at the axon ending ("synaptic knob"), small packets of a transmitter substance are released into the synaptic cleft. The transmitter can then cause a depolarization of the postsynaptic membrane of the adjacent neuron. (Modified from Hickman, C. P., and Hickman, C. P., Jr.: Biology of animals, St. Louis, 1972, The C. V. Mosby Co.)

duct an impulse not only throughout their own length but on to adjacent cells as well. This functional contact between two nerve cells is called a *synapse,* where the end of an axon connects with the dendrites or cell bodies of succeeding neurons (Fig. 3-16). The axon is actually separated from the surface of the dendrite by approximately 20 nanometers (nm). The membrane covering the knob-like end of the axon is called the *presynaptic membrane;* that of the dendrite is called the *postsynaptic membrane.* The space between the two membranes is the *synaptic cleft,* and it is across this space that the nerve impulse must be transmitted.

The direction in which an impulse is carried will be considered briefly here. Only the axon end of a neuron can transmit the nerve impulse to another neuron. Consequently, an impulse can be passed only from axon to dendrite, never from dendrite to axon. Suppose a neuron several inches long were stimulated near the middle of its length. Depolarization of the membrane would proceed toward the axon end and toward the dendrite end. When the impulse reached the dendrite, it would die out; but at the axon end the impulse could be carried on to the adjoining neuron. Each axon can release chemicals known as *transmitters.* Although several different substances are known to be transmitters, any one axon will contain only one type of transmitter. *Acetylcholine* is believed to be the most common transmitter outside the brain or spinal cord, although others (norepinephrine, serotonin, etc.) are important in various parts of the nervous system.

When an impulse arrives at a synapse, a minute amount of the transmitter is discharged into the synaptic cleft. The transmitter molecules diffuse across the cleft and alter the resting potential of the postsynaptic membrane, initiating a new impulse. It is believed that membrane permeability is changed by the transmitter, permitting ion movement through the membrane. When acetylcholine is released into the synaptic cleft, it is almost immediately destroyed by the enzyme *cholinesterase.* After the transmitter is removed the membrane quickly returns to its normal resting potential.

Structure of the nervous system. For convenience the nervous system is usually divided into three parts: the *central, peripheral,* and *autonomic* nervous systems. The central nervous system, composed of the *brain* and the *spinal cord,* is the principal integrating center of the organism and is responsible for all coordination. The peripheral nervous system, which includes the *cranial* and *spinal* nerves, carries impulses from the receptors through the central nervous system to the effectors. The peripheral system usually innervates skeletal muscles and is involved in skeletal reflex action. The autonomic nervous system includes the *sympathetic* and *parasympathetic* nervous systems. Most internal organs (heart, various glands, intestinal tract, etc.) are innervated by both sympathetic and parasympathetic fibers. If one stimulates the organ, the other usually inhibits it.

Central nervous system. Both the brain and spinal cord are made of two distinct types of tissue: *gray matter* and *white matter.* The gray matter is mainly composed of cell bodies, whereas the white matter is formed of myelinated (hence the color) fibers coursing through the brain and spinal cord. Through tracts in the white matter of the cord, sensory impulses are carried from receptors upward to the appropriate part of the brain, and motor impulses are carried downward, activating the muscles and glands.

In addition to the protection given them by the skull and vertebrae the brain and spinal cord are enveloped by three membranes, the *meninges* (Fig. 3-17). The outer sheath, the *dura mater,* is tough and composed of connective tissue. The middle layer, or *arachnoid membrane,* is so named because its delicate tissue threads resemble a spider's cobweb. The inner layer, or *pia mater,* is closely applied to the brain and spinal cord. It contains blood vessels and performs a nutritive function for this part of the nervous system. A watery substance, the *cerebrospinal fluid,* is found between the three meninges. It helps to prevent damage to the nervous tissue by providing a fluid

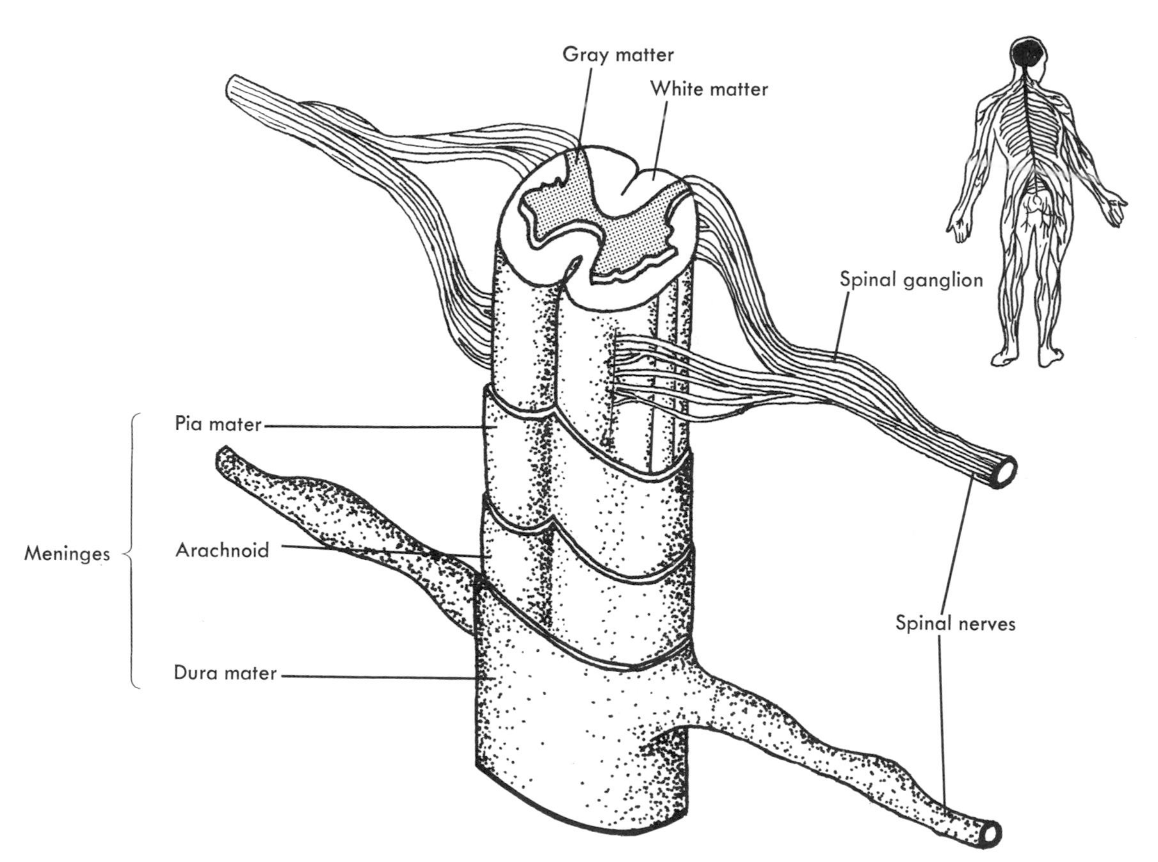

Fig. 3-17. Cross section of the spinal nerve cord. Note the arrangement of the meninges, or coverings, of the nerve cord. The spinal nerves must pass through these as they leave the spinal cord. The meninges also cover the brain. Spinal nerves leaving the spinal cord (see inset) extend to all areas of the body, with the exception of the head and the internal organs of the trunk.

cushion within the bony skull. Infection or irritation to the meningeal membranes is called *meningitis.*

The various parts of the brain, beginning with the posterior end at its junction with the spinal cord (Fig. 3-18), will be considered now. The *medulla* is basically similar to the spinal cord. It was once the most important part of the brain for all activities other than seeing, hearing, and smelling. However, in the most advanced animals the medulla has lost many of its functions to the more anterior parts of the brain. Rising above the anterior end of the medulla is the *cerebellum,* which is of extreme importance in coordination, motor activity, and posture. It varies considerably in size from one species to another, the size being generally correlated with the precision of action of the skeletal muscles. The cerebellum can, in some way, compare the sensory information from the muscles with the signal sent out from the voluntary control center in the brain. If they do not match, the cerebellum sends out a signal to the appropriate brain center and a correction is made. In this way coordinated muscle activity is accomplished. When the

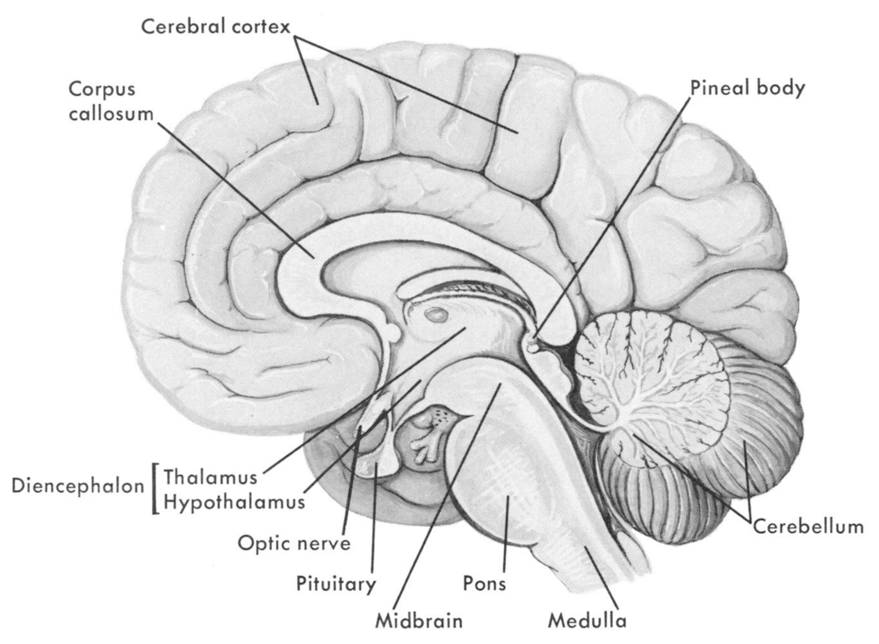

Fig. 3-18. Human brain. The left half has been removed to show the internal structure. (Modified from Schottelius, B. A., and Schottelius, D. D.: Textbook of physiology, ed. 17, St. Louis, 1973, The C. V. Mosby Co.)

cerebellum is damaged, movements become jerky and uncoordinated, much like those of an intoxicated person.

In the lower vertebrates the principal function of the *midbrain* is reception of visual signals from the eyes plus some sensory information from the ears and nose. In mammals, however, all the visual impulses are sent to the *cerebrum* rather than the midbrain, except for a few fibers that still follow the primitive pathways. The *diencephalon* includes the epithalamus, hypothalamus, and thalamus. The epithalamus is the site of the pineal body (Fig. 3-18), which is thought to be involved in the intricate and sensitive "biological clock" that in some way correlates hormonal changes to the light cycle in the environment. It may help to regulate the sex glands. The hypothalamus forms the floor and part of the side walls of the diencephalon. Such involuntary actions as temperature regulation, breathing rate, sleep, and thirst are under hypothalamic control. It sets the operating level for such things as blood pressure—one "normal" pressure when the organism is at rest and another "normal" pressure when the organism is active.

The *cerebrum*, associated with memory and learning, is poorly developed in primitive vertebrates but assumes an important role in more advanced groups. In man it is responsible for all his intellectual functions. When the cerebrum is removed from a frog, the animal can swim and hop as well as it did when intact. Even mammals such as rats and cats can still move about, although their motions may be mechanical. Their vision is impaired, but they can still see. A decerebrate human is totally blind but can still maintain breathing and swallowing for a short time before death ensues.

Peripheral nervous system. The peripheral nervous

system includes the cranial and spinal nerves. Bundles of peripheral nerve fibers are distributed to every region of the body. Those connected to the brain constitute the cranial nerves, and those connected to the spinal cord constitute the spinal nerves. The 12 pairs of cranial nerves in man (optic nerve, acoustic nerve, olfactory nerve, etc.) serve the functions of vision, hearing, smell, taste, facial sensation and muscle movement, etc.

Numerous pairs (31 pairs in man) of spinal nerves are given off at regular intervals along the length of the spinal cord (Fig. 3-17). Each spinal nerve originates in two roots, a *dorsal* and a *ventral root,* which unite to form the nerve. The dorsal root is exclusively sensory, whereas the ventral root is exclusively motor. This can be demonstrated by cutting one or the other of the roots. If the dorsal (sensory) root is severed, sensation is lost but motor activity is not affected. Conversely, severing the ventral (motor) root paralyzes the muscles served by that nerve, but it does not affect the sensory capacity of the receptors. The reflex arc, described earlier, also functions through these sensory and motor nerve fibers.

Autonomic nervous system. Internal physical adjustments are controlled primarily by the autonomic nervous system. During an emergency, digestion can wait, so that digestive activity is slowed. Along with this change, most of the blood supply is diverted from the digestive system to the skeletal muscles. Similarly, skin arterioles and capillaries constrict when the body is exposed to cold; but when the body gets too warm, they dilate and permit blood to be brought to the surface where heat is lost. These opposing effects are obtained by stimulation of one or the other of two divisions of the autonomic nervous system: the *sympathetic* and *parasympathetic* systems. Although little is known about how these divisions produce their opposing effects, at target organs the parasympathetic neurons release *acetylcholine* as their transmitter, whereas the sympathetic neurons usually release *norepinephrine.* Most internal organs are innervated by both sympathetic and parasympathetic neurons, and they generally function in opposition to each other. For example, the parasympathetic nerves stimulate the digestive tract but inhibit the heart rate; sympathetic nerves inhibit the digestive tract but stimulate the heart rate.

PART TWO

The first part of this book has discussed some broad concepts of the living world: ecosystems, some energy reactions of the cell, and some aspects of organismic physiology. This second part now examines some of the pollutants in the ecosystem and their effects on the functioning of cells and organisms.

Chapter 4 examines those metals and metal-like elements that, by their abundance and effects on plants and animals, can be regarded as pollutants. Certain of these (cadmium, lead, mercury) are known to have very serious effects; others less so. Chapters 5, 6, and 7 discuss a variety of compounds that have been associated with pollution of the environment. The division of these pollutants is somewhat arbitrary in that any one compound may reach the environment from several sources. Chapter 5 deals with inorganic and very simple organic compounds; Chapters 6 and 7 examine industrial-municipal organic compounds and agricultural organic compounds, respectively. It is hoped that the treatment of these complex compounds *by their major source* will make their relationships to one another easier to remember. Finally, in Chapter 8 a collage of miscellaneous pollutants are examined that do not allow an easy "fit" into the other chapters. Food additives, municipal sewage, and radioactive wastes are placed here.

In these chapters there is an attempt, where practical and where known, to outline the occurrence of the pollutant in the environment, its mode of entry and accumulation in the body, its symptoms of poisoning, and its mechanism of action. A "perspective" paragraph outlines the attempts made (if any) to abate the levels of a given pollutant and when such pollutants can be expected to be eliminated from our atmosphere, waters, or foods.

The physiological actions of many pollutants are not known with certainty. Much progress has been made in recent years toward pinpointing biochemical sites of action of many pollutants, but a great deal of research is still needed in this field. These chapters will show, it is hoped, what is known about these facets of pollution and also what is not known. The latter should be a challenge to us, not a source of despair.

4 METALLIC ELEMENTS AND THEIR COMPOUNDS

This chapter discusses the subjects of metals, some metal-like elements, and their toxic properties. Some of the metals in this group are familiar, for example, the lead used in fishing weights, chromium on automobile bumpers, and mercury in thermometers. People may be less acquainted with the toxic effects of some of these materials. The poisonous characteristics of some, like arsenic (a metal-like element) and lead, have been known for thousands of years, but the toxicity of others has only recently been confirmed. Beryllium and cadmium are examples of the latter group. In this chapter the more environmentally significant metals will be examined, paying particular attention to their toxic effects on man and other mammals. An attempt to assess the potential of each metal as a future environmental troublemaker will also be made.

LEAD

Description, natural occurrence, and uses. Lead (Pb) is a heavy, soft, gray metal that has been known to man for thousands of years. It occurs primarily as galena (lead sulfide) and is mined in this form in many parts of the world. Because of its low melting point and easy workability, lead has been made into a wide variety of objects and utensils. In addition, it has been used in various forms as pigments in paints and glazes for pottery. Its toxic properties have been known intermittently (the Greeks knew of them but the Romans did not) for more than 2000 years, but it remains a serious health hazard in certain situations.

Occurrence in the environment. The most widespread health hazard from lead results from its use as a pigment in paints and glazing putty (Fig. 4-1). In old buildings where paint peels off and putty comes loose from around windows, children will eat paint chips and bits of putty, thereby ingesting large quantities of lead. This is the reason for the concentration of lead poisoning cases in areas of substandard housing. Because clinical symptoms of poisoning may not appear for some time, the damage may be done before anyone is aware that the child is eating the lead-contaminated substances.

Many of the glazes used in pottery-making contain large quantities of lead salts, and eating or drinking from vessels finished with them can result in unintentional intake of large doses of lead. This is particularly true if the food is slightly acidic (e.g., fruit juices) or if it is heated in the lead-glazed containers. The Romans stored their wine in lead-glazed pottery vessels, and some historians believe that the decline of the Roman empire should be attributed to decreased mental capacity on the part of its rulers as a result of chronic lead poisoning from this source.

Large doses of lead may also be ingested from drinking "moonshine" whiskey. Most of the illicit stills in use in the United States are assembled with copper tubing soldered together with lead solder.

Fig. 4-1. Major sources of lead in the environment. The primary sources of environmental lead are automobile exhausts and smelters of copper ore and lead. The apparatus used in making moonshine whiskey and the use of lead in paints and putty are secondary sources.

Many of the larger ones employ discarded auto radiators as condensers, and these are also sources of lead. As a result, almost all samples of moonshine whiskey that have been tested reveal significant concentrations of lead. Individuals who have been regular drinkers of moonshine for many years show symptoms of chronic lead poisoning.

Lead in the atmosphere is of increasing concern to air pollution control officials. Even though relatively few cases of lead poisoning have been positively ascribed to atmospheric lead alone, the lungs provide a much easier route of entry into the body than does the digestive tract. A much greater percentage of the lead inhaled into the lungs makes its way into the system than does that entering through the intestine. The primary sources of atmospheric lead contamination are from automobile exhausts and copper and lead smelters, with the auto generating the greatest amount. Luckily, about half of the lead introduced into the air by cars falls out again within about 100 feet of roadways, but the balance of it remains suspended in the atmosphere for more extended trips, some to the farthest corners of the earth.

A great deal more is known about atmospheric lead concentrations than about the total quantities likely to be encountered in foods. A good index of average atmospheric lead can be obtained by analyzing ice laid down in the polar ice caps in particular

Fig. 4-2. Primary source of atmospheric lead. Traffic jam in Indianapolis, Indiana. Such high concentrations of automobiles in cities contribute the major amount of lead found in the atmosphere. The lead is added to gasoline for its "anti-knock" qualities and is emitted in the exhaust after the fuel is burned. (Courtesy Topics Newspapers, Inc., Indianapolis.)

years. Using this measure, atmospheric lead increased about 400% between 1750 and 1940 and another 300% between 1940 and 1965. Most of the increase occurring in the last 45 years is probably due to the advent of lead additives in automobile fuels. It has been calculated that 5.7 billion pounds of lead have been released to the air from autos since tetraethyl lead was first added to gasoline in 1923, and in the United States alone about 500 million pounds were added in 1965. The cars in the Los Angeles basin alone contribute about 30,000 pounds per day! As one would expect, atmospheric lead concentrations are higher around cities and freeways where high numbers of cars are found. Lead concentrations of 1 to 2.5 micrograms per cubic meter ($\mu g/m^3$) of air are common in cities, with peaks of 45 $\mu g/m^3$ not uncommon in areas of heavy traffic congestion (Fig. 4-2).

The discussion so far has been in terms of atmospheric lead, but as just mentioned, roughly half of the lead released by cars falls out of the air within 100 feet of roadways. This fallout contributes to high soil concentrations of lead, which is in turn absorbed by plants growing near highways. Plant lead concentrations as high as 3000 parts per million (ppm) have been measured in grasses growing near busy thoroughfares. The plants containing high lead concentrations show no signs of damage, probably because the lead in them is bound as lead phosphate, a nontoxic form. Experiments with plants indicate that in situations where available phosphorus is limited, lead has adverse effects on plant growth, which can lead to the specter of massive death of green plants whenever phosphorus decreases to low levels.

Mode of entry and accumulation in the body. As mentioned previously, the two primary routes of entry into the body are through the respiratory tract and through the walls of the digestive tract (Fig. 4-3). Only about 5% to 10% of the lead ingested ever crosses the lining of the gut to enter the circulation, but more than 40% of the lead inhaled into the lungs is absorbed into the bloodstream. We are all exposed to low concentrations of lead both in the air and in our diets. It has been estimated that the average daily intake of lead in the United States is about 300 μg,

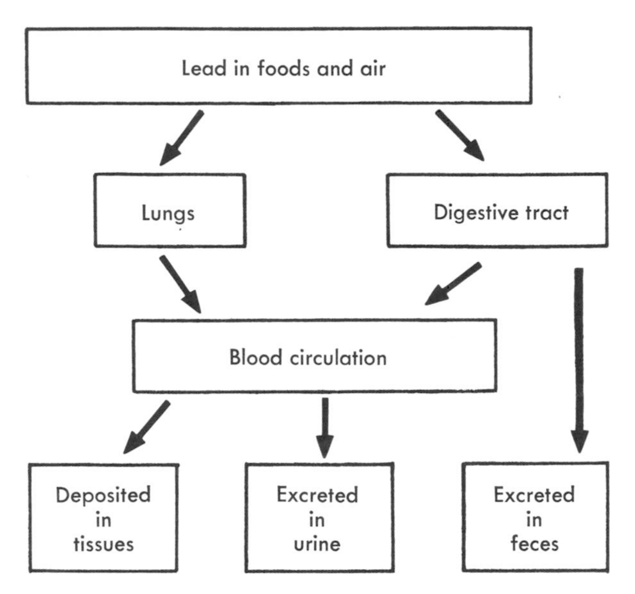

Fig. 4-3. Two primary routes by which lead enters the body. About 5% to 10% of ingested lead enters the tissues; the remainder is excreted in the feces. Of the inhaled lead, more than 40% may enter the tissues, primarily bone. Lead in the blood may be excreted in the urine.

mostly in foods and beverages. This is not enough to cause any serious long-term accumulation in the body. However, increasing the daily intake to 1000 μg is enough to cause a buildup to levels that will cause lead poisoning. Reaching this level of intake can come about either through eating more seriously contaminated foods or from breathing heavily lead-polluted air. The average daily dietary lead intake of 300 μg results in absorption of 15 to 30 μg of lead per day. Breathing air with a lead concentration of 1 $\mu g/m^3$ for 24 hours will add an additional 20 μg of lead absorbed through the lungs—the equivalent of doubling the usual dietary intake to 600 μg and getting close to the 1000 μg point, at which accumulation may result in lead poisoning. Serious lead poisoning could result from prolonged breathing of air containing 3 to 5 $\mu g/m^3$ (or more) of lead. Such concentrations might be encountered by traffic policemen, taxi drivers, and others whose jobs require working for extended periods in heavy traffic situations.

Once into the bloodstream, lead is normally either excreted by the kidneys or stored in bone. These two mechanisms prevent buildup of free lead in the general circulation and tissues of the body. When such mechanisms are overloaded by high lead intake, the excess lead builds up in the tissues, where its presence may cause difficulty.

Symptoms of poisoning. Lead has three principal sites of action in animals: the blood, kidneys, and central nervous system (Fig. 4-4). The symptoms of lead poisoning are rather nonspecific, and consequently the disease frequently progresses to an advanced stage before it is recognized. The early signs include impairment of mental function, behavior problems, and anemia. At higher levels lead in the system causes vomiting, cramps, and serious impairment of kidney and nervous system function.

The first visible result of lead poisoning is usually anemia, caused by a reduction in both the life-span and average number of red blood cells (RBC's) in the

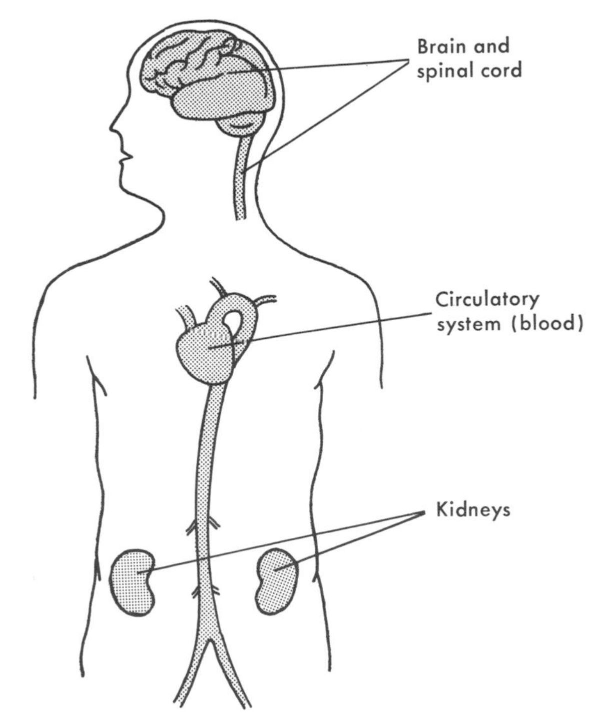

Fig. 4-4. Principal sites of action of ingested and inhaled lead.

circulating blood. This reduction in numbers of RBC's is accompanied by a reduction in the oxygen-carrying hemoglobin found in each RBC. This combination of effects leads to a serious impairment in the ability of the blood to carry oxygen from the lungs to the tissues.

Mechanisms of action. The mechanisms of lead's actions on organisms are only partially understood. Probably best known are its effects on the blood, which are produced by interference with the synthesis of heme, a major constituent of hemoglobin (Fig. 3-3). Lead has the ability to inhibit the functioning of enzymes that depend on free sulfhydryl (–SH) groups for their activity. Lead (like several other metals) ties up these sulfhydryl-active sites, making them unavailable for combining with their normal substrates. Heme synthesis involves a chain of reactions (Fig. 4-5), at least two of which require sulfhydryl-containing enzymes and hence are inactivated by elevated lead concentrations. As shown in Fig. 4-5, lead inhibits the conversion of δ-aminolevulinic acid (ALA) to porphobilinogen, and the excess ALA is excreted by the kidneys. Thus an increase in ALA in the urine provides a good confirmatory test for lead poisoning.

The effects of lead on the kidney are less completely understood than those on heme synthesis, but here, too, a pattern of interference with enzyme activity is almost certainly involved. The mitochondria of the kidney tubule cells seem particularly sensitive to increased lead concentration. The cells tend to deposit the excess lead in inclusion bodies in their nuclei, away from the sensitive mitochondria. When this mechanism fails, lead-damaged kidney tubule cells are less able to carry on their normal function of reabsorbing amino acids, glucose, and phosphates from the tubular fluid, and these substances are consequently lost in the urine. The loss of phosphate is particularly critical because it leads to a lowering of circulating phosphorus in the blood, which in turn causes calcium phosphate to be released from the bones. When this happens, lead that has been stored in the bones in a relatively innocuous form is then released into the general circulation, where it can cause difficulty. These changes in kidney tubule cells caused by acute exposure to high lead levels are largely reversible when the source of lead is removed. This is not true of other kidney effects resulting from long-term exposure to elevated lead intake. Chronic nephritis, a condition characterized by scarring and shrinking of the tissues of the kidneys, is brought about by very long-term exposures to elevated lead levels in the blood. These effects require 10 years or more to develop so that they are likely to occur in rather restricted groups: habitual moonshine whiskey drinkers represent one such group.

The third area of the body affected by exposure to high lead levels is the nervous system. These effects

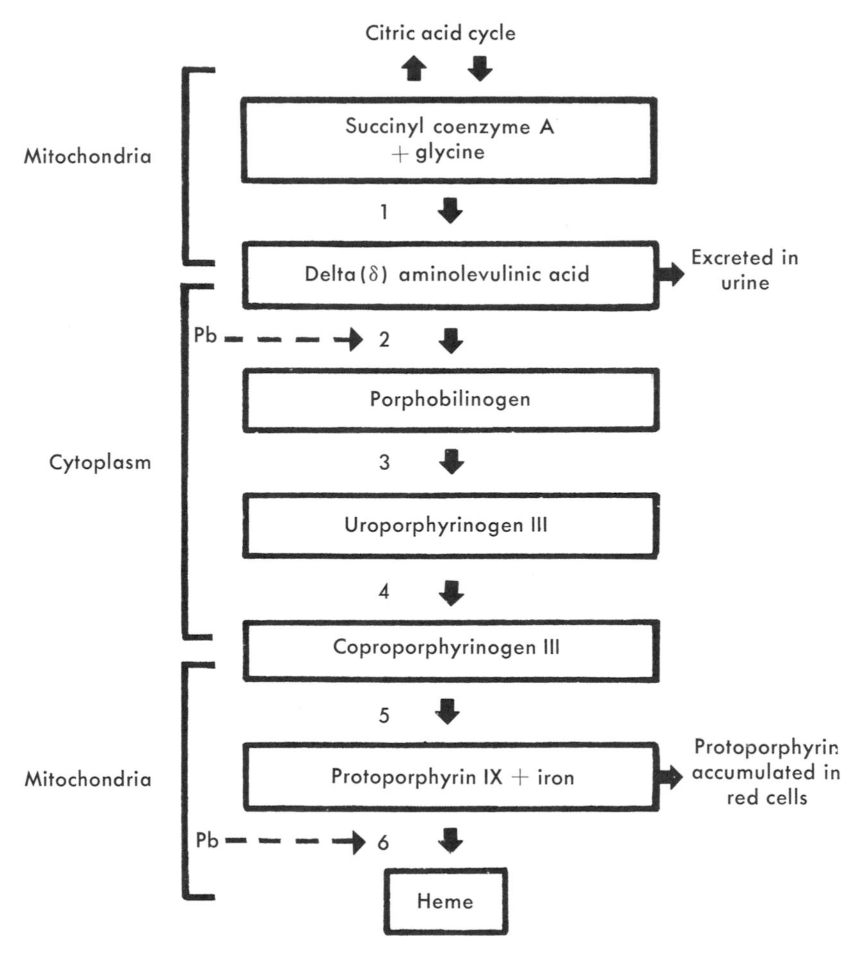

Fig. 4-5. Synthesis of heme. The raw materials for heme synthesis are succinyl-CoA (from citric acid cycle, equation on p. 31, Fig. 2-7) and glycine. These undergo a series of six reactions, each of which is catalyzed by an enzyme. At least two of these steps (indicated by broken arrows) are inhibited by lead, causing an accumulation and subsequent excretion of intermediate compounds and reduction in the amount of heme formed.

can be described here, but the underlying causes cannot be explained. The damage done to the central nervous system occurs slowly, and victims may remain unsymptomatic until brain damage is severe. The grossly detectable lead-induced changes in the brain are edema (a buildup of excess tissue fluid) and destruction of nerve cells. Edema results from an increase in permeability of the walls of the capillaries in the brain. This condition allows fluid to leak out of the capillaries and into the brain tissue, causing swelling. Since the brain is enclosed in a rigid container (the skull), swelling results in a buildup of pressure and subsequent decreases in blood flow and reduction in functional efficiency. The other effect on nervous tissue is an actual breakdown in the nerve cells. The mechanism by which this breakdown is brought about is unknown, but its consequences are far-reaching and irreversible. Together these effects are known as lead encephalopathy, and they are the most critical of the body's reactions to acute lead exposure because this form of brain damage is untreatable. Individuals who survive lead poisoning may be left as total mental defectives with no hope of return to normal activity.

Another nervous system response to long-term exposure to lead in the bloodstream is the gradual impairment of function of the motor nerves, particularly of the hands and feet. This seems to involve the Schwann cells of the myelin sheaths surrounding nerve fibers and manifests itself in an inability of the nerves to conduct impulses. When these impulses are interfered with, the hands and feet are no longer able to function adequately, and they assume a characteristic drooping posture that has been labeled "foot drop" and is a reliable indicator of chronic lead poisoning.

Perspective. The long-term outlook for reducing lead contamination of the environment may be somewhat brighter than for most of the other heavy metals because steps are already under way to remove two of the most important sources. Lead has not been widely used in interior paints since the early 1940's (when it was replaced by the cheaper titanium dioxide) so this source will probably be a thing of the past as older buildings are replaced by new ones. In the meantime, increased awareness of the potential hazard has resulted in widespread screening programs in an attempt to identify children with high levels of lead in their blood before serious permanent damage has been done. The days of tetraethyl lead in automobile fuels are also numbered, and this source of atmospheric lead will probably disappear before 1980. The reason for this is that the air pollution control devices on new cars are fouled by lead so that the cars are being built to run on lead-free gasolines. The oil companies are converting to production of lead-free fuels, and this transition will be carried out gradually as older cars requiring leaded fuels are phased out. As the sale of lead-free fuels becomes more widespread, a decline can be expected in lead concentration in the air worldwide as well as around major cities.

MERCURY

Description, natural occurrence, and uses. Mercury (Hg), or quicksilver, has long been regarded as a special material because of its unusual character as one of the few elements that are liquids at ordinary temperatures. Although a rare element comprising only 30 billionths of the earth's crust, mercury ore (cinnabar, or mercury sulfide) has been mined in Spain for twenty-seven centuries, and the toxic properties of mercury were known to the ancient Greeks. This red ore, also called vermilion, was first mined for use as a pigment, but later it was used for production of many other compounds, particularly medicines. In the Middle Ages it was used as a laxative and later for the treatment of syphilis. Today many medicines, such as antiseptics (e.g., Mercurochrome), contain certain organic compounds of mercury that can be used safely. In fact, the pure metallic form found in thermometers is not a poison, and large quantities of it can be swallowed with no apparent ill effects. However, since 1966 scientists have discovered that mercury in the environment poses a serious threat to public health. In lakes and streams, mercury can be changed from the less biologically harmful forms into lethal organic mercury compounds that are readily absorbed by the body and are poisonous to living organisms.

It is estimated that various industries currently use mercury in more than 3000 different ways (Table 4-1, Fig. 4-6). In agriculture it has become a common practice to treat wheat, rice, and other seeds with mercuric chloride as a fungicide, although various governments are beginning to restrict the use of mercury compounds. The paper and pulp industry uses mercury compounds to prevent formation of slime in the machinery of the paper mills and to

TABLE 4-1. Some mercury compounds used in industry and agriculture that enter the environment and are converted to methyl mercury

NAME	CHEMICAL FORMULA	USE
Mercuric chloride	$HgCl_2$	Treating seed potatoes; photography; steel processing; tanning leather; batteries
Phenyl mercury acetate	$C_6H_5HgOOCCH_3$	Improves storage properties of wood pulp; fungicide in water-based paint
Ethyl mercury *p*-toluene sulfonanilide	$C_{15}H_{17}HgNO_2S$	Controls smut in grain
Ethyl mercury chloride	C_2H_5HgCl	Fungicide for seed treatment
Mercuric oxide	HgO	Antifouling paint for ship bottoms
Mercuric sulfide	HgS	Coloring for plastics, sealing wax, and manufacture of fancy colored paper
Mercuric sulfate	$HgSO_4$	Batteries

improve the storage properties of pulp. In the chlorine-alkali industry mercury is used as an electrode in the preparation of chlorine for water supplies, chlorine bleaches, etc. In addition to its uses in manufacturing processes large quantities of mercury are used in lamps, batteries, and switches, which eventually find their way to rubbish heaps and incinerators.

Occurrence in the environment. During this century about 163 million pounds of mercury have been used, but little notice has been given to disposal of the waste or end product. In 1968 the chlorine-alkali producers in the United States purchased over a million pounds of mercury. Where did this mercury go? Mercury that escapes in waste water from industry and in runoff from agricultural land tends to settle in the sediment of lakes and rivers. At one time industrial discharge did not appear to pose much of a health hazard. It had been assumed that the mercury compounds settled to the stream bottom and stayed there. Much to the surprise of most scientists it has now been shown that this is frequently *not* true. Often mercury does enter the sediment as the highly insoluble sulfide and, under anaerobic (no oxygen) conditions, remains there. If, however, conditions are not completely anaerobic, the mercury may be changed through the action of microorganisms from the sulfide to the sulfate form and then into methyl mercury, which can enter the food chain as a toxic compound.

Although most natural bodies of water have mercury measuring only a few parts per billion (U. S. Public Health Service allows up to 5 parts per billion [ppb] for drinking water), much higher levels may occur even in unpolluted environments. A lake with a naturally high mercury content together with much organic sediment in the lake bed can result in the production of methyl mercury, which accumulates in aquatic organisms. Although the formation of methyl mercury and its accumulation is a natural process,

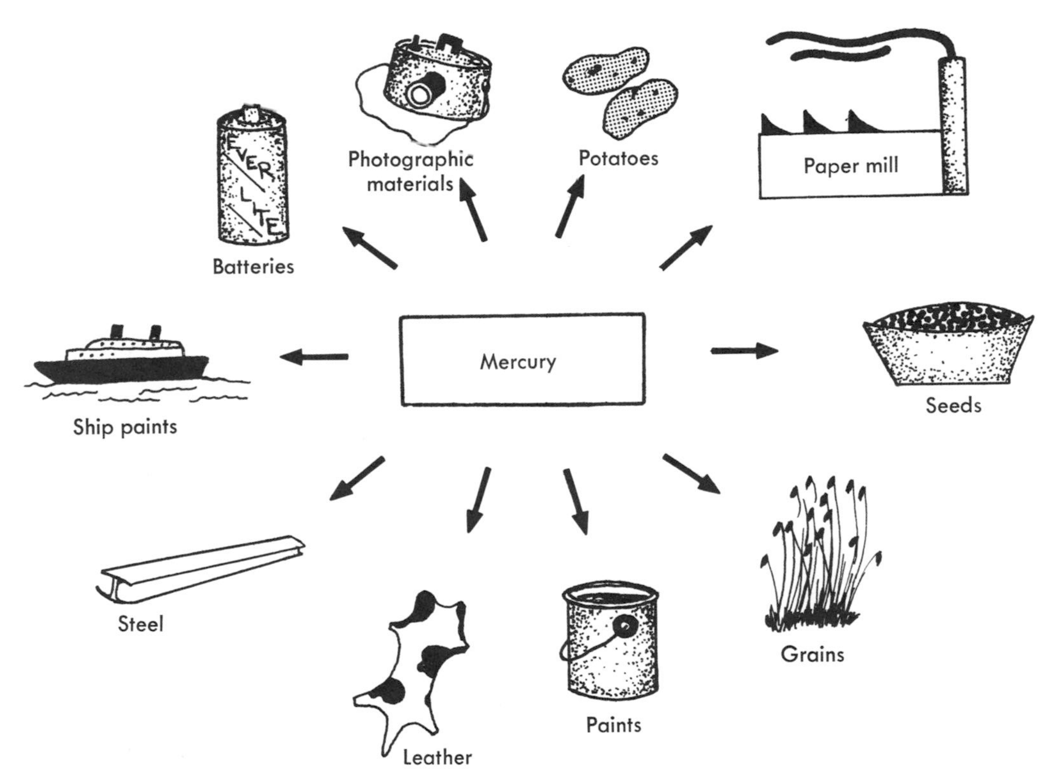

Fig. 4-6. Some uses of mercury and its compounds. The illustration indicates only a few of the more than 3000 different uses of mercury.

industrial and agricultural pollution has accelerated its production and made it more widespread. For example, in 1966 a study was made in Sweden of a pulp factory using phenyl mercury acetate. During the manufacturing process, some of the mercury was lost to the waste water. The mercury content of pike caught below the factory was five to ten times higher (up to 3.1 parts per *million*) than that in pike caught upstream.

Apparently phenyl mercury compounds (those having C_6H_5, C_6H_4, etc.) in moderate amounts do not produce toxic effects, and this is the form used in most medicines taken internally. Unfortunately these compounds are also converted into methyl and dimethyl mercury by microorganisms in rivers and lakes.

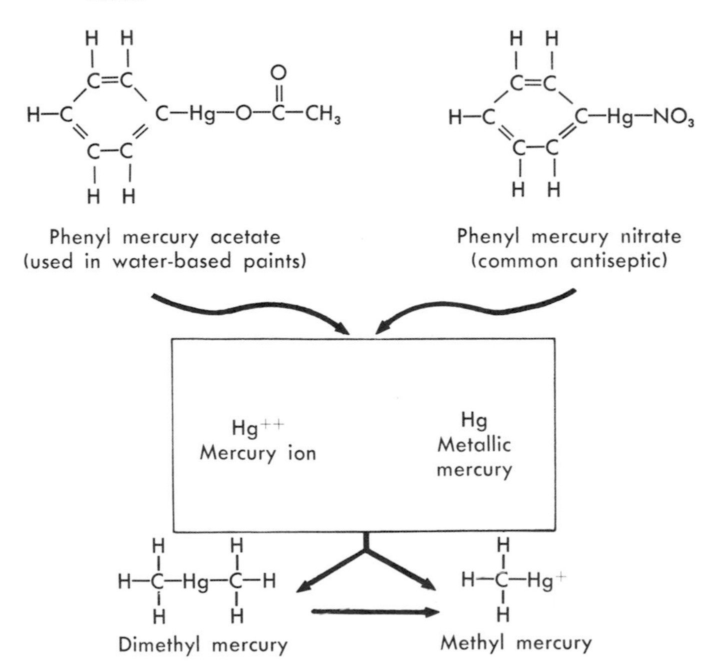

The dimethyl mercury is volatile and may escape from the lake into the atmosphere, where it is of little danger due to its low concentrations. If the body of water is at a low pH, that is, is acid, the dimethyl mercury is converted to methyl mercury. There is also evidence that the salts used in de-icing roads (calcium chloride and sodium chloride) can release mercury from bottom sediments if these salts drain into nearby lakes or rivers. It appears that the chloride ion reacts with the mercuric ion, considerably increasing the amount of soluble mercury in the water.

Mode of entry and accumulation in the body. For most people the greatest risk of mercury poisoning comes from food, although mercury can also be taken in by breathing or drinking. Since the level of mercury vapor in the air is ordinarily low, mercury poisoning via inhalation is of concern only to those who come in contact with high vapor concentrations regularly. For example, irritation and destruction of lung tissues caused by mercury vapor is an occupational hazard for workers in felt hat factories that use mercury nitrate in processing operations. Because of its low solubility, the mercury content of water is quite low. Consequently, contamination of drinking water is not currently a problem. The mercury intake from drinking water is several orders of magnitude smaller than that which would result from moderate consumption of contaminated freshwater fish.

Much of the mercury in fresh water is bound to particulate matter in the sediments of rivers and lakes. Bacteria that are principally responsible for methylation of mercury are facultative anaerobes that can live in environments with considerably lower than normal levels of oxygen. Regardless of the form of mercury that reaches them, all bacteria that can synthesize vitamin B_{12} can also carry on methyl mercury synthesis. (Vitamin B_{12} acts as a methyl transferring agent in many biologically important enzymatic reactions.) As materials, including methyl mercury, are passed up the food chain, they become more and more concentrated (Chapter 1). In aquatic food chains the mercury concentration in a carnivorous fish may be several thousand times greater than the concentration of mercury in the surrounding water. All, or nearly all, of the mercury in fish tissue is methyl mercury. In addition to acquiring mercury

TABLE 4-2. Distribution of mercury in mammals*

TISSUE	INORGANIC MERCURY COMPOUNDS	METHYL MERCURY
Blood levels (single or repeated administration)	Disappears rapidly	Marked increase, bound to red blood cells
Kidneys	100 to 1000 times blood levels	1 to 1.5 times blood levels
Liver	5 to 20 times blood levels	0.2 to 0.4 times blood levels
Brain	Equal to blood	4% to 6% of blood levels†
Half-life of single injected dose	3 to 4 days	15 days

*After Swensson, 1967; modified from Peakall, D. B., and Lovett, R. J.: Bioscience **22**(1):21, 1972.
†Due to high blood levels, total amount retained in the brain is higher than for inorganic compounds.

through food, fish can take in mercury directly through the gills. When the rate at which methyl mercury is taken up by fishes exceeds its rate of excretion, the methyl mercury accumulates (Table 4-2) and makes the fish unfit for human consumption. To further complicate matters the tissues with the greatest concentration are the ones we eat—muscle.

Because their tissue mercury level exceeded the maximum approved for human consumption, the hunting seasons for ring-necked pheasant and partridge were closed in Alberta, Canada, in 1969. Also in 1969 a check on the mercury levels of fish in the North and South Saskatchewan Rivers showed seven locations where the average values were 1 to 1.8 ppm and two other sites where the averages were 5 to 6.7 ppm. Since the maximum allowable concentration in food fish is 0.5 ppm, much of the Canadian fish supply was declared unfit for human consumption. Fish in Lake Erie were found to have up to 5 ppm. By the end of 1970 fishing was restricted in some waters of thirty-three states in the United States.

In contrast, some 1971 values for saltwater fish are 0.02 to 0.09 ppm for herring, 0.02 to 0.23 ppm for cod, 0.33 to 0.86 for tuna, and 0.82 to 1 ppm for swordfish. Although there is little doubt that the mercury level in freshwater fish has increased significantly in the last 25 years, this may not be the case with saltwater fish. Determinations of mercury in museum specimens (tuna and swordfish) show that there is no significant difference between those caught in 1879 or 1909 and those taken recently. This supports the contention that the mercury levels in wide-ranging ocean fish are not the result of man-made pollution but are of natural origin. It has been estimated that if the *total* amount of mercury processed since 1900 were put in the oceans and thoroughly mixed, it would increase the average mercury concentration of seawater (presently approximately 0.1 ppb) by no more than 1%. This suggests that although saltwater fish may accumulate mercury in excess of the 0.5 ppm (FDA maximum allowable concentration in food fish), it is a natural level resulting from their predatory nature and longevity.

Concentration of mercury through the food chain is much less marked in terrestrial than in aquatic situations. On land the concentration factor is usually on the order of two to three times normal levels, whereas it may be in the hundreds or thousands in aquatic environments. Recalling that 0.5 ppm is considered the maximum safe level, some representative values from Sweden in 1970 are milk 0.003 to 0.022 ppm, eggs 0.021 ppm, and meat 0.003 to 0.060 ppm. Studies in Canada show that in areas where mercury seed-dressing is little used the levels in seed-eating animals are low (0.1 to 0.55 ppm in Saskatchewan), but they are much higher where such usage is common (1.05 to 2.84 ppm in Alberta). Values for the United States are similar to those of other countries.

Unlike lead, mercury does not remain permanently in an organism so there is a significant drop in the mercury level when the organism leaves its contaminated environment. Table 4-3 shows the half-life (time required to eliminate one half of the mercury present) of methyl mercury in several species. Unfortunately, methyl mercury is eliminated much more slowly than are other forms of mercury. Rats eliminate 90% of their inorganic mercury in 20 days but take 150 days to remove the same proportion of methyl mercury.

Mercury is removed from the body through various routes. Birds deposit mercury in the feathers, which are regularly lost through moulting. Large amounts are also deposited in eggs. Although man gets rid of some mercury in hair, most is lost through feces and urine—which creates another problem. Because the mercury concentration in sewage sludge may be high, the use of this sludge as a fertilizer means that this mercury is available to plants and animals. Fortunately, the mercury in sewage can be removed by precipitation with aluminum sulfate.

Symptoms of poisoning. Although mercury poisoning has been recognized since early historical time, it is difficult to identify because the disease has no distinguishing characteristics. Inorganic and organic mercury compounds (except the various forms of methyl mercury) are rapidly concentrated in the kidneys and eliminated with only temporary damage. However, extended exposure can cause permanent damage or death. Methyl mercury, however, is eliminated from the body more slowly than the other forms, and the symptoms may be slower in appearing. At Minamata, Japan, release of mercury from a vinyl chloride plant into coastal waters resulted in the deaths of forty-six people between 1953 and 1960. The symptoms of methyl mercury poisoning, now called the "Minamata disease," are largely neurological. Included are lack of muscular coordination, tremors, constriction of visual fields, and difficulty in swallowing. The disease may progress to more severe symptoms, including deafness, blindness, paralysis, and kidney failure. Birds and mammals that have consumed methyl mercury develop much the same symptoms: legs of birds become lame and their neck muscles assume a contorted position, foxes run in circles, etc. Degeneration of nerve cells has been observed in chickens.

TABLE 4-3. Half-life* of methyl mercury in various organisms; data have been assembled from several sources

ORGANISM	TIME (DAYS)
Eel	900 to 1000
Pike	500 to 700
Flounder	400 to 700
Osprey	60 to 90
Man	70 to 74

*The time required to eliminate one half of the methyl mercury.

When pheasants were maintained on food containing 4.2 ppm of mercury *(ethyl mercury p-toluene sulfonanilide)*, egg production decreased by 50% to 80%. There was also a sharp increase in embryo mortality in the eggs that were laid. In humans, methyl mercury is known to accumulate in the placenta and, through "intoxification" of the fetus, causes cerebral palsy.

To further complicate matters there is no reliable diagnostic test for mercury poisoning. Urine tests are routinely used, but even the most trivial contact with mercury compounds (Mercurochrome, tooth fillings, mercury ointments, etc.) can give positive tests. Because results vary so much from day to day, tests over 6 days with results of over 0.5 mg per liter of urine should indicate mercury poisoning.

Mechanism of action. Most heavy metals interfere with the action of enzymes and other proteins that function in metabolic reactions, and this is referred to as being "poisoned." Brain cells are particularly sensitive to poisoning by methyl mercury compounds. The chemical basis for mercury's action seems to be its strong attraction to sulfur, particularly the sulfhydryl (–SH) groups in proteins. Sulfhydryl groups contribute to stabilizing the overall structure of the protein molecule. Cysteine, one of the important amino acids present in proteins, plays a major role in the three-dimensional structure of proteins by virtue of its –SH group.

```
        H
        |
H2N—C—COOH
        |
       CH2
        |
       SH
```

Cysteine

For example, the heme group of cytochrome *c* in the electron transport system (Chapter 2) is linked to the protein portion by the sulfhydryl groups of three cysteines, and these links can be broken by heavy metals such as mercury.

Research on yeast cells shows that mercury acts on the sulfhydryl groups of cell membrane proteins and upsets the normal permeability. It appears that the mercury is not acting on the metabolism of the cell but on the physical structure of the membrane. Other research indicates that mercury can directly affect active transport in the cell. It can inhibit uptake of glucose in the intestine or of sodium in the kidney. The primary action of mercury on the kidney seems to be a depression of the active transport mechanisms involving sodium, potassium, and certain anions. Mercury can also induce chromosomal changes. Studies have shown a direct correlation between chromosomal breaks and mercury concentration. Low concentrations may also inactivate the mitotic spindle or cause mitosis to stop. In human white blood cells methyl mercury inhibits formation of the mitotic spindle at concentrations as low as 0.25 ppm. The chromosomes double but remain at metaphase, a condition known as "metaphase arrest." Other mercury compounds will produce similar results.

Since sulfhydryl groups in any protein can be inactivated by mercury, almost every aspect of metabolism is susceptible to its action.

Perspective. The U. S. Environmental Protection Agency in 1972 banned interstate shipment of many mercury-based pesticides, including the entire class of alkyl mercury compounds (of which methyl mercury is one). Also suspended was the application of all other mercury compounds to rice seed, laundry fabrics, and antifouling paint for boats. Similar action was taken in Sweden in 1965, and the decreased use of methyl mercury did not affect crop yields.

Although many industries have stopped discharging mercury into streams and lakes, there are abundant deposits of inorganic mercury already present in sediments of these bodies of water, and it is believed that these mercurial residues can persist up to 100 years in polluted lakes.

Since 1970 it has become clear that concentrations of mercurial compounds below the proposed water-quality standards can have detrimental effects on photosynthesis by phytoplankton. In Japan industrial waste water may by law contain up to 10 ppb of methyl mercury, yet it has been shown that in freshwater plankton 1 ppb can cause a significant reduction in photosynthesis. At 50 ppb (just five times the allowable limit) essentially all uptake of carbon is stopped, and there is a complete inhibition of plankton growth.

Tuna and swordfish have a tendency to accumulate mercury in excess of the 0.5 ppm limit allowed in food for human consumption. Yet, when Japanese quail were given a diet containing 20 ppm of methyl mercury, they survived longer when the mercury was given in tuna meat than when they received the same amount of methyl mercury in a corn-soya diet. Something in the tuna meat diet prolonged survival in quail that were given high concentrations of mercury. Tuna normally has a relatively high concentration of selenium, and it also tends to accumulate more selenium when mercury is present. Thus it appears that tuna accumulates mercury and selenium together. It has been suggested that selenium in tuna reduces the toxicity of mercury by complexing (chemically combining) with it in the blood, thereby decreasing the availability of each element. Interestingly, both mercury and selenium are present in about the same concentrations in the oceans (Se = 0.09 ppb; Hg = 0.1 ppb). This indicates that the danger of eating tuna may be less than was originally assumed. However, too little is known about this problem at present to relax food safety standards, especially as they apply to freshwater fish. Industrial pollution has made the accumulation of mercury in freshwater fish a chronic hazard.

CADMIUM

Description, natural occurrence, and uses. Cadmium (Cd) is a soft, silver-white metal related in its chemical properties to zinc and mercury. There are no cadmium ores as such; rather, the metal occurs as a constituent of zinc, copper, and lead ores, usually in the form of cadmium sulfide (CdS) and cadmium carbonate ($CdCO_3$). The metal also forms the oxide, CdO, and the sulfate, $CdSO_4$, and the oxide may react with water to form the hydroxide, $Cd(OH)_2$. About half (3150 tons) of all the cadmium consumed in the United States each year is used by the electroplating industry. This process coats objects with cadmium, using the metal as the anode (positive terminal) in an electrolytic bath, which passes direct current to the object (as the negative terminal) to be coated.

The nickel-cadmium battery industry used some 200 tons of cadmium in 1968 in the manufacture of negative terminal plates. Cadmium is also used as a stabilizer that adds life to plastics, as a pigment in plastics and paint, in the manufacture of metal alloys, photographic supplies, and glass, and in rubber curing. The metal also finds use in agricultural fungicides and is a common impurity in phosphate fertilizers.

Occurrence in the environment. The presence of large amounts of cadmium in the atmosphere and waters is viewed with alarm, since many observations have established a highly probable causal relationship between the metal and a variety of physiological disorders. Cadmium as an environmental pollutant is a product of industrialization. It is released to the environment as a by-product at two levels: (1) the processing of cadmium-bearing ores and reclaiming of scrap metals containing cadmium and (2) the disposal, usually by incineration, of manufactured products containing cadmium. The metal is released as a fine dust or mist.

TABLE 4-4. Cadmium emission to the atmosphere of the United States, 1968 (estimated)*

SOURCES OF EMISSION	ESTIMATED QUANTITY EMITTED (TONS)
Refining of Cd-bearing ores	1050.0
Reclaiming scrap steel	1000.0
Copper reclaiming from auto radiators	125.0
Incineration of Cd-bearing wastes	95.0
Production of Cd paint pigments	10.05
Wear of vehicular tires	5.07
Production of plastic pigments, stabilizers	3.0
Manufacture of Cd-bearing alloys	2.5
Burning of vehicular motor oil	0.91
Miscellaneous manufacturing	0.56
	2292.09

*From National Air Pollution Control Administration report, February, 1970, U. S. Department of Health, Education and Welfare, Raleigh, N. C.

In 1968 the National Air Pollution Control Administration (NAPCA) in a study of cadmium emission to the atmosphere of the United States reported that some 4.6 million pounds (2300 tons) of the metal was released, most as a by-product of either the processing of cadmium-bearing ores or in the disposal of cadmium-containing products. (Table 4-4 lists some of the sources of cadmium emission and approximate amounts of each source.) The report cited in Table 4-4 shows that about 89% of atmospheric cadmium is due to ore refining and processing and to the reclamation of scrap metal containing cadmium.

A survey of cadmium in United States waterways has not been made. A few analyses at selected sites, however, indicate that some waters are carrying an increased burden of the metal. In a study of rivers and reservoirs by the U. S. Geological Survey, reported in 1970, 4% of 720 samples showed a cadmium content in excess of the 0.01 ppm set by the U. S. Public Health Service as the maximum allowable level for drinking water. Two thirds of the highly contaminated samples came from waters that were downstream from large cities or industrial complexes. Another study published the same year showed that in a few rare cases the cadmium levels in water at the customer's faucet exceeded the U. S. Public Health Service's upper limit.

The use of cadmium in fungicides and its presence in fertilizers draw immediate attention to the levels of the metal in foods. Presumably, cadmium could enter plants as a natural soil constituent or could enter from the application of fungicides and fertilizers. Studies have shown no consistent correlation, how-

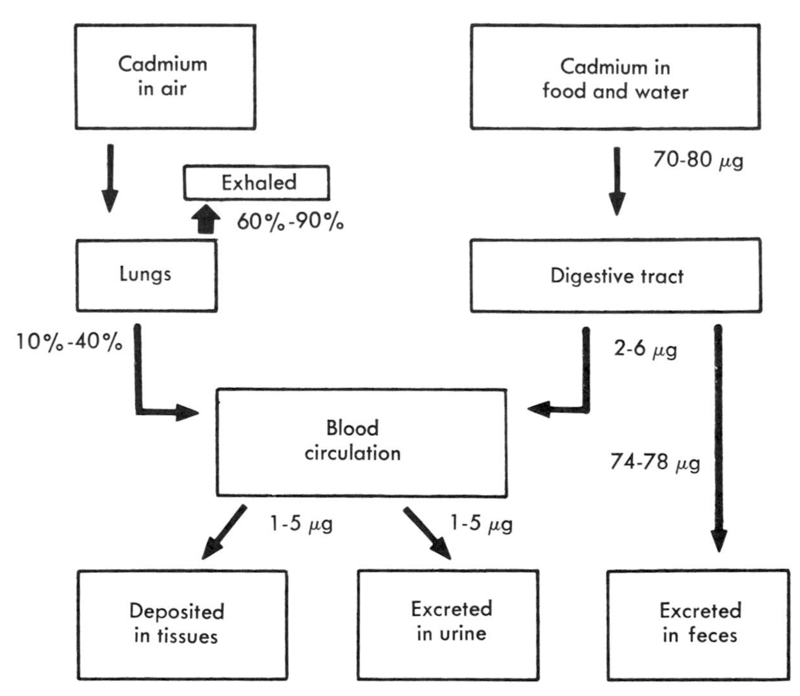

Fig. 4-7. Fate of ingested and inhaled cadmium. Up to 97% of the ingested cadmium is excreted in the feces and does not accumulate in the body. (μg = microgram.)

ever, between the use of cadmium-containing fungicides and fertilizers and the levels of cadmium in plants and soils. Some plants apparently can concentrate cadmium from the soil or from fungicides; others cannot.

Mode of entry and accumulation in the body. Cadmium may gain entrance to the body in food or water and is then absorbed through the intestinal wall into the blood (Fig. 4-7). One study indicates that the daily intake per person may be from 70 to 80 µg. The study suggested that most of this amount (up to 97%) is excreted in the feces without being absorbed. The remainder, from 2 to 6 µg, is absorbed into the system. Although cadmium ingestion can be a significant route of poisoning, numerous studies in the United States and in other industrialized nations show the most common route to be the inhaling of cadmium dust from industrial operations. The metal gains direct access to the blood through the alveolar wall of the lungs. According to a Swedish study, a higher percentage of inhaled cadmium (10% to 40%) is absorbed through the lungs than through the intestinal wall (3% to 8%). Not all the cadmium absorbed is accumulated in the tissues; 5 µg or less may be excreted daily in the urine. A "normal" atmospheric concentration of cadmium is about 0.001 $\mu g/m^3$. The atmospheric load near cadmium industries has been measured at 0.1 to 0.5 $\mu g/m^3$. A pack of twenty cigarettes smoked in one 24-hour period may add 2 to 4 µg of cadmium to the daily intake. The amount of cadmium absorbed from food or water appears to be related to the amounts of certain nutrients in the diet. Persons whose diets are deficient in protein and calcium may absorb as much as 10% of the cadmium ingested. There appears also to be an inverse relation between zinc (an essential micronutrient) and cadmium absorption: those who have a low zinc intake tend to absorb higher quantities of cadmium than if their dietary zinc is at normal levels.

Studies in animals and humans show that absorbed cadmium may be deposited in a variety of tissues

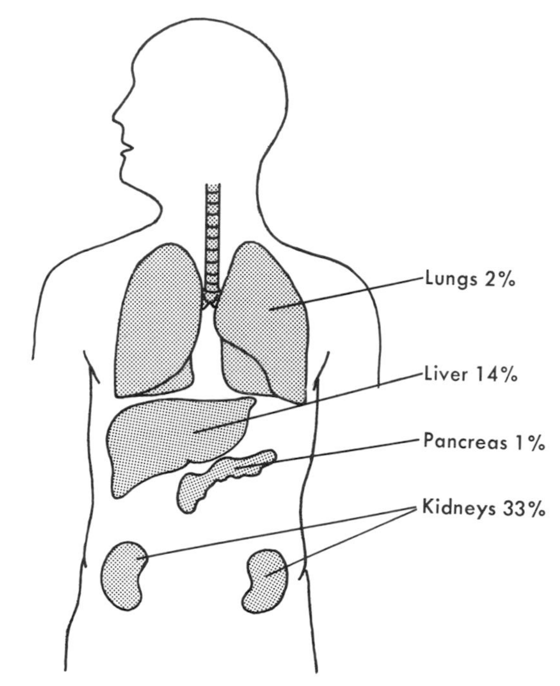

Fig. 4-8. Major sites of cadmium deposition in the body. The kidneys and liver together accumulate almost half of all the cadmium stored in the body, with the balance being widely distributed among the other tissues.

(Fig. 4-8), but two organs stand out as cadmium concentrators: the kidneys and the liver. The kidney cortex (outer layer) may hold 33% of the entire cadmium load, the liver 14%, the lungs 2%, and the pancreas less than 1%. The remaining 50% is distributed among the other tissues. Over 90% of the cadmium in the blood is found in the red cells, probably bound to hemoglobin, the respiratory pigment. Laboratory experiments have shown that such cadmium-laden red cells may be engulfed and destroyed by certain white blood cells.

Symptoms of poisoning. Symptoms related to chronic cadmium poisoning are numerous but tend to fall into a consistent pattern. Factory workers who

TABLE 4-5. Comparison of atmospheric cadmium and hypertension mortality ratios in selected United States cities

CITY	ATMOSPHERIC CADMIUM ($\mu g/m^3$)	HYPERTENSIVE MORTALITY RATIO (AVERAGE RATIO = 100)
Chattanooga, Tenn.	0.001	87.6
Cincinnati, Ohio	0.007	111.2
Scranton, Pa.	0.011	135.2
New York, N. Y.	0.013	115.3
Newark, N. J.	0.018	119.3
Indianapolis, Ind.	0.019	102.5
Philadelphia, Pa.	0.023	127.6
Chicago, Ill.	0.062	123.8

have inhaled cadmium dust for several years report general tiredness, nervousness, dryness of the mouth, impaired sense of smell, and shortness of breath (dyspnea). Other symptoms are sore throat, chest cramps, pain in the small of the back, and poor appetite leading to loss of weight and yellowing of the teeth. Perhaps the most characteristic findings in cases of long-term cadmium poisoning are pulmonary emphysema, low molecular weight proteins in the urine, cloudy urine, and cirrhosis of the liver. Excess quantities of amino acids have also been noted in the urine.

By far the most suggestive correlation recorded to date is the relation between cadmium and hypertension (high blood pressure). Among the best evidence of this relationship is a study reported in 1966 that compared atmospheric cadmium in several cities with death rates due to hypertension and arteriosclerotic heart disease. Using a base-line value of 100 as the standard mortality ratio for hypertension and heart disease, Table 4-5 compares the atmospheric cadmium load to this standard mortality ratio in a few of the twenty-eight cities surveyed. The comparisons shown in Table 4-5 are suggestive of a causal relationship between cadmium and hypertension, but they do not afford clear proof of a cause-and-effect relationship.

Some later (1970) studies on this effect of cadmium have shown no correlation. One complicating factor is that hypertensive diseases may likely be related not only to cadmium levels but also to a zinc deficiency. The rationale here is that cadmium and zinc, due to their chemical similarities, may be in competition with one another in the biochemistry of the body tissues.

Injections of cadmium compounds in animals (CdS, CdO, $CdSO_4$) have initiated the growth of tumors (cancerous growths) at the injection site. Some studies in humans appear to show a similar relation also, but a direct causal relation has not been shown. Cadmium oxide dust has been shown to be associated with cancer of the prostate gland.

Perhaps the most bizarre and pathetic record of long-term, heavy cadmium poisoning comes from a mining community in northern Japan. The disease is characterized by degeneration of the bones and is known as *Itai-Itai* (variously translated as "It Hurts-It Hurts," or "Ouch-Ouch") disease. The bones decalcify, leaving them soft, and they fracture under slight pressure. Although high levels of lead were present in its victims (perhaps contributing to the bone softening), the concentration of cadmium was extraordinary (over 11,000 ppm in rib bones). Clinical symptoms appeared mostly in females over 40 years of age who had several children. Cadmium was the major, if not the only, cause of the disease. Two researchers in an article on cadmium poisoning, published in 1957, wrote: "Cadmium has probably more lethal possibilities than any of the other metals." Since that date, some 15 years ago, the warning is becoming more and more significant.

Mechanism of action. Research into the mechanism of action of cadmium in cells and tissues indicates the charged cadmium ion (Cd^{+2}) is the active form of the metal rather than any of its compounds. The more soluble forms of cadmium tend to be incorporated into the tissues at higher rates than the less soluble (CdS, CdO) forms. These salts can be solubilized, however, by a reaction with slightly acid or alkaline water, as may exist in the intestinal tract or the lungs.

Having reached the various body organs as Cd^{+2}, the metal is bound (in the kidneys and liver) by a protein called metallothionein. This protein contains many sulfhydryl groups in its structure and appears to act as an agent that effectively deactivates cadmium ions in the tissues. Injections of cadmium stimulated the synthesis of the protein, and the tentative conclusion is that these sulfhydryl compounds act as a defense mechanism, protecting the tissue from cadmium poisoning. A further conclusion is that this defense breaks down if the load of cadmium becomes excessive. The excess cadmium then becomes toxic.

The biochemical site of action of cadmium in cells has been fairly clearly pinpointed (Fig. 4-9). Cadmium binds to the membranes of mitochondria, those small intracellular organelles in which most of the body's energy-generating reactions take place (Chapter 2). It will be recalled that two major energy pathways exist in the mitochondria: (1) the citric acid cycle, which generates nicotinamide and flavin compounds containing electrons and hydrogen atoms from nutrient molecules, and (2) the electron transport, or respiratory, chain, which conducts these electrons and hydrogen atoms to molecular oxygen,

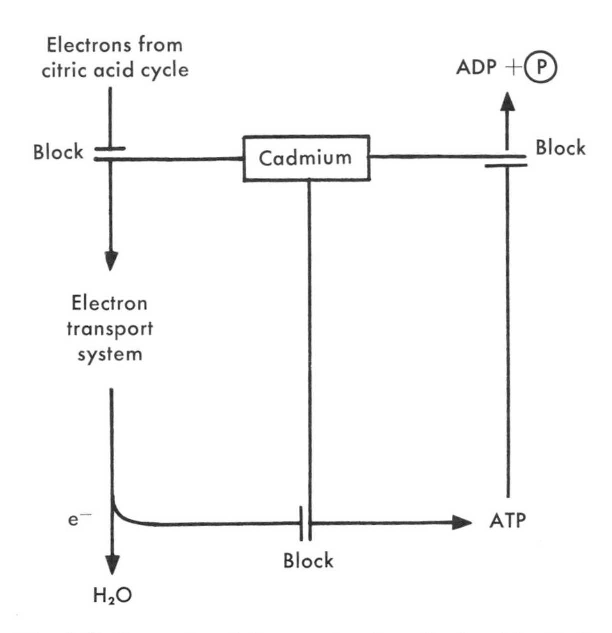

Fig. 4-9. Sites of cadmium action in mitochondria. Cadmium acts on enzymes of the citric acid cycle and on the electron transport chain located on mitochondrial membranes. The metal (1) blocks the transfer of electrons from the citric acid cycle to the electron transport chain, (2) blocks the synthesis of ATP, and also (3) blocks the breakdown of ATP to ADP.

forming water. This hydrogen and electron conduction results in the synthesis of adenosine triphosphate (ATP), the ubiquitous, high-energy compound required in almost all synthetic reactions in the body. The mitochondrial electron transport chain is nearly the body's sole source of ATP.

Cadmium ion is seen to be active in at least three places in the cell: (1) it binds to sulfhydryl groups of enzymes necessary for the transfer of electrons from the nicotinamide and flavin compounds from the citric acid cycle to the compounds of the electron transport chain; (2) cadmium ion also binds to and inactivates one or more enzymes necessary for the synthesis of ATP by the respiratory chain; and (3) cadmium ion also binds to the enzyme adenosine triphosphatase (ATPase), required to split off a phosphate from ATP. Without this reaction ATP cannot function as a source of necessary energy in cellular reactions. Some workers suggest that cadmium may inhibit the function of enzymes dependent on copper, cobalt, or zinc.

The action of cadmium ion therefore seems to be where it hurts most: the inhibition of enzymes required in energy-generating reactions. The observation that long exposure to cadmium decreases phosphate in blood serum may be related to the tissues' inability to split ATP. The fact that low serum phosphate results in a loss of bound minerals may be related to the dread Japanese Itai-Itai disease.

Perspective. Many authors believe that when detected, it is already too late to combat cadmium poisoning. This attitude is based in part on the finding that when elevated cadmium levels appear in the urine, indicating cadmium poisoning, the level of the metal is so high that the kidneys have been damaged beyond repair and nothing can be done to restore the patient to health. Since cadmium is accumulated by the body, it is imperative that human exposure levels be reduced wherever the metal exceeds allowable concentration limits. This requires continued surveillance of potential cadmium pollution sources, their avoidance by humans, and their correction. It is possible to block the poisoning action of cadmium in tissues with certain chemicals that bind the cadmium ion, thus protecting essential enzymes in the energy reaction pathways. So far this has been done only on isolated tissues and cells in the laboratory. No extensive human trials have been conducted, but where they have been done, these chemicals (ethylene diamine tetraacetate [EDTA] and British anti-Lewisite [BAL], a sulfhydryl compound) have been shown to alleviate the symptoms of acute cadmium poisoning. Vitamin D treatment is also reported to result in improvement. There is, therefore, hope that the acute form of cadmium poisoning can be effectively treated.

BERYLLIUM

Description, natural occurrence, and uses. Beryllium (Be), although one of the least known environmental pollutants, is important because it is one of the most toxic nonradioactive elements known. Beryllium is a light, stiff metal with a high melting point, properties that are bringing about increasing uses of beryllium in modern technology. It is commonly used as a component in alloys, particularly of copper. In nature, beryllium exists in a variety of ores, but the only one that is of commercial importance is beryl (beryllium aluminum silicate). Beryl is seldom mined alone, but it is recovered as a by-product of mica or feldspar refining operations.

Occurrence in the environment. One of the major uses of beryllium in the past has been in the fluorescent light industry. Beryllium oxide was used as a major ingredient in the phosphorus coating the inside of the tubes. Although this practice was discontinued in 1949, it was the exposure of workers in this industry that brought beryllium disease (berylliosis) to the attention of the general public. Beryllium contamination of the environment is largely confined to industrial plants that refine it from its ore and those that use it in alloying or machining. The air in the vicinity of such plants may contain significant quantities of beryllium compounds. Beryllium is also a minor (0.1 to 31 ppm) contaminant of most coals, and it is probable that most of the airborne beryllium in areas remote from processing plants comes from burning this fuel.

Mode of entry and accumulation in the body. Beryllium gains access to the body primarily through the lungs, since almost none is absorbed through the wall of the digestive tract (Fig. 4-10). Repeated or prolonged contact of the skin with beryllium compounds may result in localized effects (dermatitis), but the quantities of beryllium absorbed by this route are usually small. There is no known mechanism for excretion of beryllium, and consequently, once in the

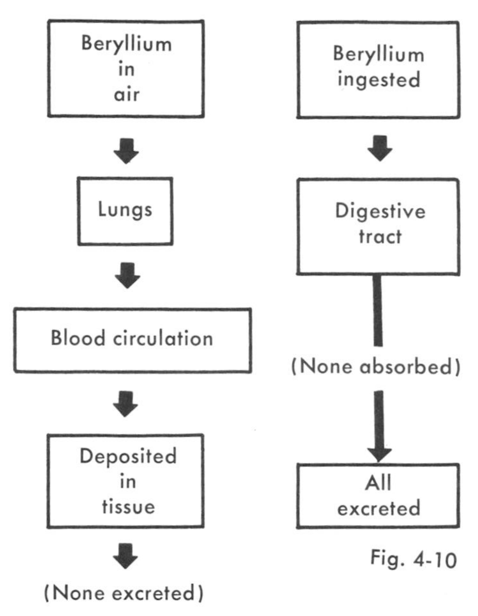

Fig. 4-10. Entrance into, and accumulation of, beryllium in the body. All ingested beryllium is excreted. All inhaled beryllium is accumulated in body tissues, primarily tissues of the respiratory tract.

system it remains and accumulates as long as the individual lives.

Symptoms of poisoning. Beryllium poisoning occurs in both acute and chronic forms, depending on the intensity and duration of exposure. The acute form results from relatively short exposures to high concentrations of beryllium or its compounds. It is characterized by inflammation of the entire respiratory tract, bronchitis, chemical pneumonitis, edema (buildup of fluids) of the lungs, coughing, shortness of breath, weakness, fatigue, weight loss, and anemia. Beryllium particles in the lungs frequently induce the formation of granular lesions that are visible on x-ray examination and interfere with breathing. If exposure has not been too severe, recovery from the acute form of berylliosis may be complete in several weeks. Single high-intensity exposures have proved fatal in several instances, however. In one experiment using large numbers of animals over 10% of those exposed to beryllium salts contracted lung tumors. Most of the remaining 90% developed a variety of pulmonary lesions and abnormal cell growth of the alveoli (air sacs).

The chronic form of berylliosis is more insidious than the acute form, sometimes developing many years (up to 30) after exposure has ceased. Its early symptoms are usually similar to those of the acute form: coughing, shortness of breath, weight loss, etc. Chronic berylliosis progresses to a degeneration of most body functions and, ultimately, death. The right side of the heart may become enlarged as a result of increasing pulmonary insufficiency. There are considerable individual differences in sensitivity to beryllium and hence rather scattered patterns of occurrence of berylliosis in the vicinity of sources. Individuals living as far as ¾ mile from a processing plant have contracted the disease, whereas many workers in the plant who were exposed to beryllium concentrations hundreds of times higher were unaffected. This so-called "neighborhood effect" has led to the hypothesis that beryllium poisoning may depend on an inherited flaw in the individual's metabolic machinery.

Mechanism of action. Little is known about how beryllium causes damage in its victims. The variations in sensitivity just referred to have led to an investigation of the immune response, and findings indicate that the antigen-antibody reactions of the body are involved. Nitrogen metabolism is upset, as evidenced by elevated levels of proteins in the blood. One study has implicated beryllium compounds in inhibition of alkaline phosphatase activity. Alkaline phosphatase is an enzyme that catalyzes reactions, making free phosphate available by hydrolyzing organic phosphorus compounds. This free phosphate is required by many reactions (e.g., reaction on p. 28, Fig. 2-6). Alkaline phosphatase is also required for calcium phosphate deposition in bone.

Perspective. United States air quality standards for

beryllium have been set at 0.01 μg/m³ for ambient air, 2 μg/m³ for industrial exposure (averaged over an 8 hour working day), and 25 μg/m³ for maximum short-term exposure. The industrial users of beryllium have taken steps to meet these standards, and it is hoped that with their attainment beryllium poisoning will become a disease of the past.

NICKEL

Description, natural occurrence, and uses. Nickel (Ni) is a hard gray-white metal, ferromagnetic like iron, and highly resistant to oxidation. There are almost no pure nickel deposits in the world that can be worked economically. Nickel occurs most commonly in oil and coal deposits and with iron and copper ores as nickel sulfide (NiS), the oxide (NiO), the silicate ($NiSO_3 \cdot xH_2O$), and the arsenide (NiAs). Consumption of nickel in the United States in 1966 totaled 187,833 short tons out of a total world production of 475,000 tons. In 1966 about 82% of the United States consumption of nickel was in the manufacture of stainless and other heat-resistant steels, the production of nickel alloys, alloy steel manufacture, and as anodes in the electroplating industry. Nickel-aluminum compounds are also used as catalysts in hydrogenation and dehydrogenation of organic compounds, bleaching, drying of oils, water purification, and catalytic combustion of organic compounds in the exhaust of internal combustion engines. Nickel carbonyl ($Ni[CO]_4$) is also used as a catalyst. It is generated as a short-lived reactant in the refining of nickel from other metal ores, or it is prepared from nickel chloride for specific use as a catalyst. Nickel carbonyl is extremely toxic. Several nickel organic compounds have been synthesized as fuel additives, purportedly reducing wear on engine parts, improving lubrication, and reducing carbon deposits.

Occurrence in the environment. Asbestos, coal, and crude oil have been shown to contain nickel. The burning of coal and petroleum products appears to be the principal source of atmospheric nickel. The nickel content of coal shows a wide variation, depending on the source, from 0 to 42 ppm. Nonweathered coal is reported to contain as much as 95 ppm nickel. Again depending on the source, the nickel concentration in coal ash may be as high as 1%.

Nickel in crude oil has been measured at 55 and 110 ppm in two studies. The emissions from diesel engine exhaust varied widely in one study (500 to 10,000 μg per gram of particulate matter), depending on engine speed and load conditions.

Nickel has been sampled from the atmosphere of over 100 cities for several years. In 1964 the average concentration at these sites was 0.032 μg/m³, and the maximum (East Chicago, Indiana) was 0.69 μg/m³. Incinerator emissions also add to atmospheric nickel. Analysis of particulate emissions from one municipal incinerator showed an ash content as high as 10% nickel (100,000 ppm). The highly toxic nickel carbonyl is formed whenever nickel comes in contact with carbon monoxide at appropriate temperatures (50° C or even at 25° C, which is slightly above room temperature, when the monoxide concentration exceeds 100 ppm). The burning of coal and oil, gasoline additives, and incineration of nickel-containing products therefore may be sources of nickel carbonyl. The American Industrial Hygiene Association in 1968 recommended an 8-hour exposure limit for industrial workers not to exceed 7 μg/m³ (0.001 ppm) for nickel carbonyl and 1000 μg/m³ for nickel metal and soluble nickel compounds.

Modes of entry and accumulation in the body. Atmospheric nickel and its compounds may enter the respiratory tract or be absorbed through the skin (Fig. 4-11). The various salts of the metal are toxic, particularly the carbonyl. Ingested nickel metal is excreted rapidly in the feces, whereas inhaled nickel appears in the urine. The metal has been found in food and water and in animal tissues, but in levels of only a few micrograms per gram. Plants appear to have a low tolerance to nickel, and so vegetable foods are not likely to be a significant source.

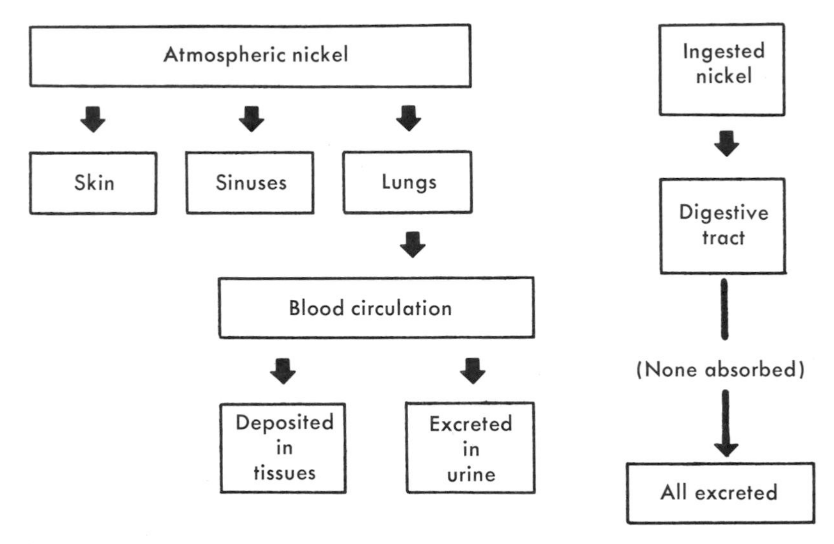

Fig. 4-11. Entrance into, and accumulation of, nickel in the body. All ingested nickel is excreted in the feces, but inhaled nickel may be deposited in the tissues or excreted in the urine. Nickel is deposited in liver, kidney, heart, brain, intestine, and skin. It is not retained in the human lung.

Nickel is not retained in the human lung but can be found consistently in the skin. Cigarette smoke contains the highly toxic carbonyl of nickel (1.6 to 3 μg of nickel per cigarette). Experiments with rats show that the injected metal is rapidly eliminated from the body. After 3 days nickel was detectable only in the liver, adrenal glands, lungs, eyes, spleen, and teeth. Experiments that exposed mice to nickel carbonyl indicate that accumulation of nickel is proportional to the frequency of exposure. In these experiments nickel was found in the microgram range (maximum of 11.37 μg per gram of liver tissue) in liver, lung, kidney, heart, brain, and intestine.

Symptoms of poisoning. Pure nickel metal is relatively nontoxic. Contact with nickel in nickel-plated articles can result in a dermatitis (inflammation of the skin). Numerous instances are reported among workers in nickel refineries and in nickel electroplating plants. Inhaling of nickel fumes and of nickel carbonyl has been shown to produce severe respiratory ailments in humans and in experimental animals. The alveolar (air sac) cells of the lungs become enlarged, cell division rate is increased, cells proliferate, and hemorrhages occur. Initial symptoms of severe exposure to the carbonyl may be dizziness and headache, followed by vomiting and shortness of breath. Delayed symptoms may include cyanosis (blue tinge of skin and fingernails due to lack of oxygen), elevated body temperature, and delirium indicating involvement of the central nervous system. Extreme poisoning may result in death within 1 week.

A well-documented effect of exposure to nickel fumes and to nickel carbonyl is a significant increase in nasal and lung cancer. In humans the evidence that nickel and its compounds is the causative agent is still circumstantial, but very strong. A study of one nickel refinery showed that death from nasal cancer among workers was 150 times the expected rate, and for lung cancer, five times. In another study the death rate from sinus cancer was 200 times more than expected. In the cases examined the workers had been exposed to nickel dust and fumes for several years. The dust emitted from one nickel refinery was shown to contain the sulfate, the sulfide, and the oxide of nickel at levels totaling in excess of 85% of the total metals and moisture in the dust. Other compounds included the oxides of copper, cobalt, iron, and silicon.

The carcinogenicity of nickel compounds in experimental animals presents even stronger evidence. Tolerance of the metal alone is quite high in most animals, but nickel salts are highly toxic when injected directly into the bloodstream. In some studies a single exposure to nickel carbonyl of 35 to 80 ppm produced cancer. The LD_{50} is a measure of lethality due to some given quantity of a toxic substance. It is the dose of a substance that is lethal to 50% of the animals in an experimental group. The LD_{50} for nickel carbonyl inhalation for 30 minutes was shown to be 10 ppm for mice, 34.7 ppm for rats, and 274 ppm for cats.

Mechanism of action. As noted earlier, the presence of nickel compounds in the lungs results in significant alterations in the alveolar cells. These changes include clumping of nucleoprotein, increases in nuclear and cytoplasmic RNA content, and possibly increase of protein also. Both acute and chronic exposure of rats to inhaled nickel carbonyl showed that the nickel was located principally in the endoplasmic reticulum of the lungs and liver. After prolonged exposure the nickel was found in the cell nuclei and mitochondria also. Experiments with a carcinogenic compound, benzo[*a*]pyrene (a five-ring hydrocarbon), have shown that nickel carbonyl (inhaled or injected into rats) suppresses the action of an enzyme, benzo[*a*]pyrene hydroxylase, which specifically acts on the hydrocarbon. It has been suggested that the resulting enzyme inhibition (possibly inhibition of its synthesis) allows retention of the hydrocarbon by cells, promoting carcinogenesis. Other studies have demonstrated that dimercaprol or BAL (British anti-Lewisite) and EDTA (ethylene diamine tetraacetate—a metal binding agent) alleviated the symptoms of nickel carbonyl poisoning. The protection afforded by BAL indicates that nickel compounds may act by binding to sulfur atoms of enzymes.

Perspective. The environmental air standard upper limit for an 8-hour exposure has been recommended not to exceed 7 $\mu g/m^3$. The maximum concentration measured in 1964 was about one tenth of this amount in the air over a heavily industrialized city. The many well-documented cases of nasal and lung cancer in persons who work in industrial operations concerned specifically with nickel indicate that the allowable limit is either too high or that some industries are not meeting the standards. Except for the specific cases involving refining and nickel electroplating, atmospheric nickel does not appear to be a general health problem at present. Emissions of nickel as fumes or dust can be controlled by filtering, precipitation, and scrubbing procedures already available and in use. The dangerous nickel carbonyl can be decomposed to nickel and carbon monoxide by heating above 60° C (140° F). This compound therefore need present no problem of disposal.

ARSENIC

Description, natural occurrence, and uses. Arsenic (As) has properties of both a metal and a nonmetal. The element occurs naturally in coal (up to 16 μg per gram) and in oil. It is a ubiquitous element but usually occurs in very small concentrations (a few parts per million in soil, up to 100 parts per *billion* in sea water). It can be concentrated by some marine and freshwater animals (40 ppm in marine shrimp and bass). Because of its widespread (though low) occurrence, arsenic can be found in most foods.

There are only a few deposits of the pure element; arsenic usually is found as the sulfide (As_2S_3, As_2S_5) and compounded with oxygen as arsenides, arsenates, and arsenites in lead, copper, and gold ores.

Arsenic-containing pesticides and herbicides form the main use of arsenic. In 1964 about 14.5 million pounds of arsenic-containing herbicides, insecticides,

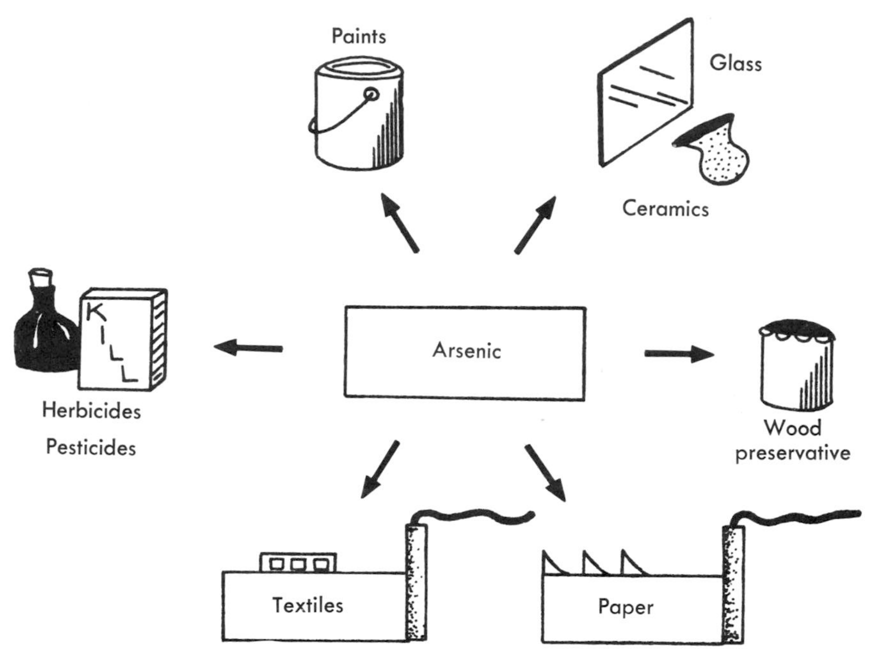

Fig. 4-12. Some important uses of arsenic compounds. The major use of arsenic has been in pesticides, herbicides, and defoliants. It also is used as a dehydrating agent for cotton.

and defoliants were used in the United States. About 5 million pounds of this total was in the form of arsenic acid, which was used as a dehydrating agent for cotton. The use of arsenic in insecticides and herbicides has declined as organic pesticides such as DDT and 2,4-D have developed. Arsenic compounds are also found in paints, glass, ceramics, and wood preservatives, and they are used in the textile, tanning, and paper industries (Fig. 4-12).

Occurrence in the environment. Large quantities of arsenic and its compounds are produced from the smelting of lead, copper, zinc, and gold ores, and from some cadmium and nickel ores. Among the most toxic of the effluents from these operations is the very volatile white arsenic (arsenic trioxide, As_2O_3). This compound is removed from exhaust gases chemically or by condensation. These processes are not 100% efficient and so may provide a significant source of air pollution. For example, 48,000 tons of zinc ore that contain 0.07% arsenic could yield up to 34 tons of arsenic in the gases from the ore-smelting operations. Due to the use of arsenic acid (H_3AsO_4) as a desiccant for cotton, arsenic is present in the dust emitted from cotton gins. It has been estimated that a bale of cotton may contain from 0.03 to over 6 grams of arsenic. Coal also contains arsenic, and wherever coal is burned, arsenic will be emitted to the atmosphere. The air over New York City averaged 0.03 $\mu g/m^3$ of arsenic in 1964. An 8-hour exposure limit to arsenic and its compounds not to exceed 500 $\mu g/m^3$ of arsenic has been recommended for industrial workers in the United States.

Mode of entry and accumulation in the body. Compounds of arsenic may be absorbed by the body by inhalation, ingestion, and absorption through the

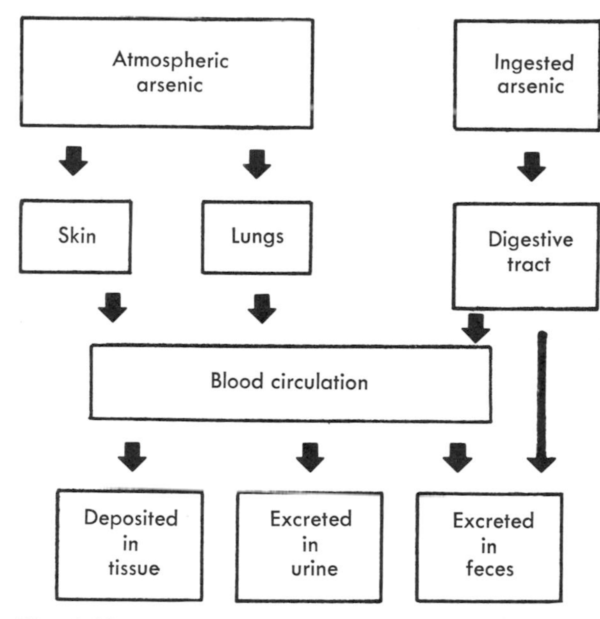

Fig. 4-13. Entrance into, and accumulation of, arsenic in the body. Virtually all tissues accumulate arsenic, especially hair and nails. The body can rid itslef of about half of its arsenic load every 30 to 60 hours.

skin (Fig. 4-13). Arsenic is detectable in all body tissues. It is excreted primarily in the urine. The body can rid itself of about half the arsenic load every 30 to 60 hours. Arsenic is commonly found in keratin (horny) structures such as hair and nails. Arsenic is carried in the blood bound to the protein part of hemoglobin.

Symptoms of poisoning. The severity of arsenic poisoning depends on the concentration and type of compound. Most arsenic compounds are extremely toxic. The element itself is nontoxic. Acute poisoning results in gastrointestinal inflammation, nausea and vomiting, diarrhea, feeble and irregular heartbeat (resulting in part from heart muscle necrosis), coma, and possible death in a few days. Arsine gas (AsH_3) may be fatal in dilutions of 50 ppm. Chronic arsenic poisoning is somewhat less severe, resulting in muscle weakness, loss of appetite, occasional vomiting, gastrointestinal pains, constipation, inflammation of nasal and oral mucous membranes, coughing, skin lesions (on skin contact), and a graying of the skin. Death from arsenic poisoning is due to respiratory and circulatory failure.

There has been a great deal of controversy over the carcinogenic action of arsenic compounds. Early reports indicated that cancerous growths were due to arsenic exposure, but later investigations have not confirmed these reports. The difficulty may lie in distinguishing between two different chemical states of arsenic. The arsenate (pentavalent form, such as sodium arsenate, Na_2HAsO_4) is less toxic than the highly poisonous arsenite (trivalent form, such as sodium arsenite, $NaAsO_2$). Arsenic in the very dangerous arsine gas (AsH_3) is in the trivalent form. It is therefore not warranted at this time to place airborne arsenic compounds among the list of carcinogens.

Mechanism of action. Arsenic combines readily with sulfur groups of enzymes, thus preventing or inhibiting their catalytic function. In one well-documented case the phosphate group that is normally added to glyceraldehyde phosphate in glycolysis (reaction no. 5 on p. 28) can be replaced by an arsenate group ($HAsO_4^{-2}$). The arsenate substitution follows the oxidation of the aldehyde to the acid, and the resulting compound is arsenophosphoglyceric acid. This compound is unstable, and the arsenate is immediately hydrolyzed off with water to yield the free arsenate and phosphoglyceric acid, as follows:

$$\text{Gal}—\text{Ⓟ} + HAsO_4^{-2} \xrightarrow{NADH_2} AsO_4—\text{Gac}—\text{Ⓟ} \xrightarrow{H_2O} HAsO_4^{-2} + \text{Gac}—\text{Ⓟ}$$

Ordinarily, using phosphate instead of arsenate, the glyceric acid generated would carry *two* phosphates. These are ultimately combined with ADP to yield two ATP. Because of the involvement of arsenate, the acid carries only *one* phosphate, and so only one ATP is subsequently generated. If arsenate inhibits the critical ATP-generating enzymes of the electron transport chain (Fig. 2-8) or the citric acid cycle enzymes (Fig. 2-7), almost all ATP generation is shut down. Although it is not known if all enzymes are susceptible to attack by arsenic (all enzymes do contain sulfur), it is easy to understand now why arsenic compounds are so deadly to animals and plants.

Perspective. Arsenic trioxide is reported to be fatal in doses ranging from 70,000 to 180,000 μg (a few thousandths of an ounce). Arsine gas is considered hazardous at levels of 210,000 μg/m³. The average concentration of arsenic in the air of 133 cities and towns in 1964 was 0.02 μg/m³. These atmospheric levels therefore do not appear to present a health hazard to the general population. Workers in ore smelting operations, however, may be exposed occasionally to much higher concentrations. Methods currently used to remove particulate material from dust and fumes (if the temperature is not much higher than 100° C) are sufficient for the removal and entrapment of arsenic and its compounds. Fabric filters, electrostatic precipitators, condensation by cooling, and wet "scrubbing" by vacuum pumps are all in use and are effective, but some smelters have yet to be equipped with such control devices.

OTHER ELEMENTS

A number of other elements of varying toxicity have been detected in the environment, primarily as air pollutants (Fig. 4-14). They probably occur also in the nation's waterways, but likely at low concentrations. Of this number, antimony, bismuth, and tin are relatively toxic. No natural physiological function for antimony or bismuth is known.

Antimony (Sb). Antimony is frequently used in

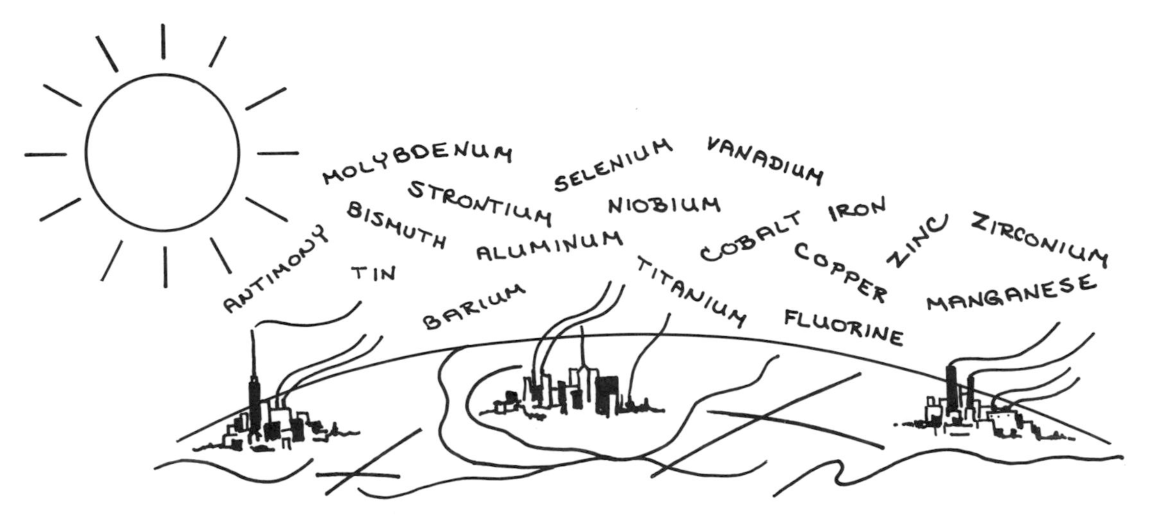

Fig. 4-14. Atmospheric elements with pollution potential. Over two dozen metals and metal-like elements have been detected in the atmosphere. Some are extremely hazardous.

the manufacture of alloys, batteries, medicinals, and in the textile industry. It is present occasionally in small amounts in the air and is not regarded as a general or industrial health hazard. Its symptoms of poisoning and effects are similar to those of arsenic. The lethal dose to humans is reported to be between 100 and 200 mg. There is no evidence at present that antimony is related to cancer. Its arsenic-like effects indicate that its mechanism of action may be the inhibition of sulfur-containing enzymes.

Bismuth (Bi). Bismuth compounds are generally of low toxicity. The metal is found in coal and so is emitted to the air on burning of this fuel. Bismuth poisoning results in kidney and liver damage, but only when present in large amounts. The symptoms are similar to those of lead and arsenic poisoning and are relieved by BAL treatment.

Tin (Sn). Tin finds use primarily in the plating industry. Acute poisoning results in vomiting, diarrhea, constipation, weight loss, and anemia. Organic forms of tin are much more toxic than the inorganic forms. Organic tin compounds enter the nervous system and accumulate in heart tissue. Atmospheric tin is present in very low levels and is absent from the air of many cities. It does not present a health hazard. Recently tin has been shown to be essential for normal growth of rats.

• • •

A number of elements and their compounds occur in the air but at very low, generally nontoxic levels. These include aluminum, barium, niobium, strontium, titanium, and zirconium. These elements appear to have no normal physiological function.

Aluminum (Al). Aluminum is a natural pollutant. It is found in all tissues but accumulates in the lungs. It is ingested daily in the food and is not toxic at these levels. It does not present a health hazard. Aluminum has not been shown to be essential to normal body functions.

Barium (Ba). Barium is also a natural pollutant. Like aluminum it is present at very low levels in the air and is ingested daily in the food. It accumulates in the lungs as the insoluble sulfate. The soluble salts of barium are moderately toxic, and large doses may be fatal. It has no known function in the body.

Niobium (Nb). Niobium is found in petroleum and so, presumably, is in the air when oil products are burned. Its toxicity is low; it does not cause tumors when ingested by rats. The element is not known to be essential to body functions.

Strontium (Sr). Strontium is ingested daily in food and so is a natural pollutant. Its present levels are not regarded as toxic. The dangerous radioactive form of strontium, Sr-90, will be discussed in a later section. Strontium has not been demonstrated to be essential to life, although its chemical similarity to calcium allows it to replace this element in some tissues.

Titanium (Ti). Titanium metal and its oxide are nontoxic, but titanium tetrachloride has caused trouble to titanium workers. Contact of the tetrachloride with water releases heat and yields the corrosive hydrochloric acid. Body areas exposed to the tetrachloride of titanium should be *wiped* dry before being flushed with water. Titanium is present in the air in concentrations less than 0.03 $\mu g/m^3$. It accumulates in skin and lungs but has not demonstrated toxicity when fed to mice for life. This metal therefore appears to present no general health hazard, and it has not been shown to be essential in the diet.

Zirconium (Zr). Zirconium is detectable in automobile engine exhausts, indicating its presence in gasoline or oil. The metal accumulates in the lungs. It does not appear to be toxic when fed to rats. Its very low atmospheric levels pose no health problem.

• • •

Several additional elements that are essential to life have been found in (or are probably in) the air. These elements are required in minute (trace) amounts in metabolism by the cell, usually as cofactors with enzymes. They are chromium, cobalt, copper, fluorine, iron, manganese, molybdenum, vanadium, selenium, and zinc.

Chromium (Cr). Chromic acid, chromates, and bichromates are used in electroplating, steel making, tanning, photography, and chemical synthesis. Although the metal presents no general health hazard, industrial exposure to large quantities has produced serious effects, ranging from skin lesions and gastrointestinal ulcers to lung cancer. The lethal dose of chromate (as potassium chromate) is about 5 grams. Chromium is a necessary cofactor with the hormone insulin in promoting glucose utilization.

Cobalt (Co). Cobalt is found in most tissues at levels of about 0.3 μg per gram of tissue. Cobalt dust from refinery and alloy plants has caused dermatitis, gastrointestinal pain, vomiting, and low blood pressure. Large doses of cobalt as the chloride result in goiter (enlargement of the thyroid gland) and polycythemia (high red blood cell count). The lethal dose is estimated to be between 50 and 500 mg per kilogram of body weight (about 0.3 to 3 ounces per 165 pounds of body weight). The oxides and the more toxic sulfides of cobalt have caused cancer in rats (but not in mice). Air levels are low and present no general hazard. Cobalt is a constituent of vitamin B_{12} and its derivatives. It acts in coenzymes that are required in fatty acid degradation, in the metabolism of some amino acids, in red blood cell formation, in methyl ($-CH_3$) transfer reactions, and in DNA synthesis.

Copper (Cu). Small amounts of copper are found in the air, probably from ore smelting operations. Beside its wide use as a malleable metal, copper is used in herbicides, fungicides, and rat-killing compounds. Inhalation of copper or its salts has produced nasal congestion, and large doses can induce vomiting. Liver and kidney damage result from massive absorption of copper, which could lead to anemia and circulatory failure. Neither copper nor its salts is known to produce tumors. The element is a natural and essential constituent of cells. It is a cofactor in at least one of the enzymes of the electron transport chain, cytochrome oxidase (Cyt a+a_3, Fig. 2-8), in an enzyme required in the metabolism of the amino acid tyrosine, in the respiratory pigment of many invertebrates, and in perhaps ten other enzymes including one required for iron utilization.

Fluorine (F). Fluorine is a highly reactive gas found in coal at levels of about 80 ppm and in polluted air from 0.01 to 0.4 $\mu g/m^3$. Fluorine is emitted during the manufacture of steel, aluminum, glass, bricks, and ceramics. It can be a seriously harmful pollutant in localized areas. Hydrofluoric acid (HF), which is extremely corrosive, causes skin ulceration on contact and can burn and destroy mucous membranes if inhaled. The effects of the fluoride ion (F^-) vary with its concentration in the body. Trace amounts can check the occurrence of dental caries, but large amounts result in pitting and brown mottling of the tooth surface. Experiments indicate that fluorine is involved in calcium and phosphorus metabolism, two elements required for the synthesis of bone. Its toxic effects derive from the inhibition of enzyme action, notably inhibition of the enzyme (enolase) that dehydrates phosphoglyceric acid to form phosphoenolpyruvic acid in glycolysis (reaction on p. 28). Fluoride also inhibits enzymes in the metabolism of the fatty acids. The recommended concentration in drinking water is about 1 ppm.

Iron (Fe). Iron is an essential element that is present in the air and in water in small amounts, probably from ore smelting and scrap reclaiming operations. Most of the body content of iron is derived from the diet and drinking water. Iron is present in all tissues as a constituent of the respiratory pigments hemoglobin and myoglobin, in the cytochromes of the electron transport chain, and in several other enzymes (peroxidase, catalase, and a dehydrogenase enzyme in the citric acid cycle). There is no mechanism for the excretion of iron so the element accumulates in the body. Large doses of iron and its oxides have caused hemorrhaging and necrosis of stomach and intestinal tissue, liver necrosis, and pulmonary congestion. Circulatory collapse, coma, and death may result within 24 hours. The sulfides and oxides of iron have not been shown to be carcinogenic. Its low concentration in air is not regarded as a health hazard.

Manganese (Mn). Low levels of manganese are also found in the air and present no health problem, generally. Manganese poisoning has been reported, however, from workers in manganese ore plants, from manganese grinders, and from manganese alloy workers. Poisoning is caused by the various oxides of manganese and results in bronchitis, pneumonia, headache, lethargy, muscle weakness, tremors, and mental deterioration. BAL is reported to be useful in treatment of manganese oxide poisoning. The element occurs as a necessary constituent in several enzyme systems, including those that transfer phosphate groups between compounds, in urea formation, and in an enzyme involved in the metabolism of pyruvic acid (pyruvate carboxylase).

Molybdenum (Mo). Molybdenum is an element that also occurs in small amounts in the air. It accumulates primarily in the liver and kidneys. It is not retained by the lungs. It is a necessary constituent of a few oxidase enzymes involved in nucleic acid metabolism and in the utilization of the element nitrogen.

Selenium (Se). Selenium is a nonmetallic element found associated with sulfur in coal and petroleum in amounts of less than 7 ppm in the former. Selenium is found in the soil and vegetation, and consumption of vegetation by grazing animals has caused "loco," or alkali disease in such animals. The disease can be fatal to animals but is uncommon in humans. Selenium alone may be toxic as a dust in fumes; its soluble salts are much more toxic, however, causing lesions of the liver, kidneys, heart, lungs, spleen, stomach, and intestines. In large doses selenium is carcinogenic in rats. Some observations indicate that an organic form of the element may be essential in animals in trace amounts (2 to 4 μg per 100 grams of diet food). Exclusion of selenium from the diet of lambs resulted in muscular dystrophy. Injections of selenium plus vitamin E prevent the condition from developing. The element appears to act in the electron transport chain of mitochondria. Selenium poisoning is reported to result from the inhibition of sulfur enzymes, thus being similar in large concentrations to the action of arsenic. A relationship between selenium and mercury in tuna has been demonstrated. Mercury poisoning seems to be attenuated in the presence of selenium (p. 82).

Vanadium (V). Vanadium compounds find use in steel making and are present in petroleum and fuel oil. The resulting fumes and smoke from burning oil may contain significant quantities of the oxide of vanadium. Vanadium accumulates in the lungs and is generally of low toxicity when ingested. Some 90% of workers in a vanadium refinery were reported to suffer from bronchitis, some seriously. Vitamin C (ascorbic acid) has protected experimental animals from death by vanadium poisoning. Environmental levels of vanadium are very low and offer no general health hazard. Present evidence indicates that vanadium is normally involved in the mineralization of bones and teeth. Its absence from the diet slows the growth rate of rats.

Zinc (Zn). Zinc is an essential element that is present in the air and can be found in all body tissues. Its source in air is primarily from refining, alloying, welding, and cutting of metals containing zinc. The element itself is not toxic, but its salts (oxide and chloride) have caused vomiting, diarrhea, and gastric pain. Massive exposure has resulted in death from pneumonia and shock. Zinc is only slowly absorbed by the body and does not accumulate in the lungs. Most of the body's source of zinc is dietary. Zinc is a necessary constituent of enzymes involved in dehydrogenation and protein degradation, in alcohol metabolism, and in carbon dioxide formation.

• • •

All these elements discussed here are present in minor or trace quantities in the air and do not present a general health hazard. Since most are poisonous at some concentration, their levels should be monitored so atmospheric loads can be followed. A great deal more research is required to determine the biochemical effects of these elements. The site of action of only a very few is known with certainty.

SUGGESTED READING

Lead

Bazell, R. J.: Lead poisoning: Zoo animals may be the first victims, Science **173**:130-131, 1971.

Chisolm, J. J., Jr.: Lead poisoning, Scientific American **224**(2):15-23, 1971.

Mercury

Goldwater, L. J.: Mercury in the environment, Scientific American **224**(5):15-21, 1971.

Passow, H., Rothstein, A., and Clarkson, T. W.: The general pharmacology of the heavy metals, Pharmacological Reviews **13**:185-224, 1961.

Peakall, D. B., and Lovett, R. J.: Mercury: its occurrence and effects in the ecosystem, Bioscience **22**(1):20-25, 1972.

Wood, J. M.: A progress report on mercury, Environment **14**(1):33-39, 1972.

Cadmium

Athanassiadis, Y. C.: Preliminary air pollution survey of cadmium and its compounds: A literature review. Prepared for U. S. Department of Health, Education and Welfare, Public Health Service, Raleigh, N. C., 1969, National Air Pollution Control Administration.

Friberg, L.: Health hazards in the manufacture of alkaline accumulators with special reference to chronic cadmium poisoning, Acta Medica Scandinavica **138**(suppl. 240):1-124, 1950.

McGaull, J.: Building a shorter life, Environment **13**(7):3-41, 1971.

Mustafa, M. G., and Cross, C. E.: Pulmonary alveolar macrophage: Oxidative metabolism of isolated cells and mitochondria and effect of cadmium ion on electron- and energy-transfer reactions, Biochemistry **10**(23):4176-4185, 1971.

Mustafa, M. G., Peterson, P. A., Munn, R. J., and Cross, C. E.: Effects of cadmium ion on metabolism of lung cells. In Englund, H. M., and Beery, W. T., editors: Proceedings of the Second International Clean Air Congress, New York, 1971, Academic Press, Inc.

Schroeder, H. A., Balassa, J. J., and Vinton, W. H., Jr.: Chromium, cadmium and lead in rats: Effects on life span, tumors, and tissue levels, Journal of Nutrition **86**(1):51-66, 1965.

Beryllium

Durocher, N. L.: Preliminary air pollution survey of beryllium and its compounds, Raleigh, N. C., 1969, National Air Pollution Control Administration.

Knapp, C. E.: Beryllium—hazardous air pollutant, Environmental Science and Technology **5**:584-585, 1971.

Nickel

Mastromatteo, E.: Nickel: A review of its occupational health aspects, Journal of Occupational Medicine **9**(3):127-136, 1967.

Sullivan, R. J.: Preliminary air pollution survey of nickel and its compounds: A literature review. Prepared for U. S. Department of Health, Education and Welfare, Public Health Service, Raleigh, N. C., 1969, National Air Pollution Control Administration.

Sunderman, F. W., Jr.: Inhibition of induction of benzpyrene hydroxylase by nickel carbonyl, Cancer Research **27**(5):950-955, 1967.

U. S. Department of Health, Education, and Welfare, Public Health Service, Division of Air Pollution: Air quality data from the national air sampling network and contributing state and local networks, 1964-1965, Washington, D. C., 1966, Government Printing Office.

Arsenic

Birmingham, D. J., Key, M. M., Holaday, D. A., and Paone, V. B.: An outbreak of arsenical dermatosis in a mining community, Archives of Dermatology **91**:457-464, 1965.

Schroeder, H. A.: A sensible look at air pollution by metals, Archives of Environmental Health **21**(6):798-806, 1970.

Schroeder, H. A.: Metals in the air, Environment **13**(8):18-32, 1971.

Sullivan, R. J.: Preliminary air pollution survey of arsenic and its compounds: A literature review. Prepared for U. S. Department of Health, Education and Welfare, Public Health Service, Division of Air Pollution, Raleigh, N. C., 1966, U. S. Department of Health, Education and Welfare.

Other metals

Arena, J. M.: Poisoning, Springfield, Ill., 1963, Charles C Thomas, Publisher.

Frieden, E.: The chemical elements of life, Scientific American **227**(7):52-60, 1972.

Schroeder, H. A.: A sensible look at air pollution by metals, Archives of Environmental Health **21**(6):798-806, 1970.

Schroeder, H. A.: Metals in the air, Environment **13**(8):18-32, 1971.

The following references are individual publications, each covering a different metal. Each is titled: "Preliminary Air Pollution Survey: a Literature Review," and was prepared for the U. S. Department of Health, Education, and Welfare, Public Health Service, Consumer Protection and Environmental Health Service, National Air Pollution Control Administration, Raleigh, N. C. All are dated October, 1969.

Miner, S.: Barium and its compounds.
Sullivan, R. J.: Chromium and its compounds.
Sullivan, R. J.: Iron and its compounds.
Sullivan, R. J.: Manganese and its compounds.
Stahl, Q. R.: Selenium and its compounds.
Athanassiadis, Y. C.: Vanadium and its compounds.
Athanassiadis, Y. C.: Zinc and its compounds.

5 INORGANIC AND SIMPLE ORGANIC COMPOUNDS

From all the publicity that has been given to pollution in recent years we get the idea that all of the harmful materials in our environment are man-made. Not so! Every pollutant discussed in this chapter is naturally occurring and is a normal part of the environment. The pollution problem arises when individuals start to use these naturally occurring materials in large quantities. Such quantities cannot move through the normal biogeochemical cycles fast enough because of some environmental factor. They start to accumulate in the air or water or food—this is pollution. In a nutshell, pollution is merely too much of some material in one place.

Smoke belching from a factory smokestack was once a welcome sight. However, as more and more factories were built with an ever-increasing amount of smoke, there came a time when these materials began to remain in man's environment and form a part of the atmosphere he breathed. All the components of smoke (NO_2, SO_2, CO, CO_2, unburned hydrocarbons, etc.) are part of the atmosphere. Farmers add phosphate and nitrate fertilizers to their crops just as nature normally does, but they do it in such quantities that ultimately these materials also enter drinking water. Just because a material is present in the natural environment does not mean that it is not harmful. As with most things, a little is essential, a moderate amount is beneficial, but a lot can do you harm. This chapter is concerned with those substances present in nature that are essential and beneficial, but if handled improperly, can do harm to us and our world.

ASBESTOS

Description. Inhalation of asbestos dust has long been recognized as the cause of a condition known as asbestosis, but only recently have the mechanisms by which asbestos affects the body tissues been found. Present evidence indicates that asbestos pollution is a far more serious problem than was previously thought.

Asbestos is the general name for a variety of minerals that can easily be unraveled into fibers. Asbestos minerals occur in nearly every country in the world, but only a few are commercially valuable. By far the most important of the asbestos minerals is chrysotile, or hydrated magnesium silicate ($3MgO - 2SiO_2 \cdot 2H_2O \cdot H_2O$), making up about 90% of the world production. The most important deposits of chrysotile in the United States are in California, Arizona, and Vermont. Another form of asbestos called crocidolite is mined in North Carolina. However, these deposits do not meet present demands so that most of the country's asbestos is imported from Canada and Africa.

The value of asbestos comes from the indestructible nature of products made with the mineral (Fig. 5-1). It will not burn and thus has many uses as a heat-resistant structural material or for thermal insulation.

Sources of pollution. It is difficult to determine the level of asbestos pollution because the asbestos particles in the atmosphere are frequently masked by the other dusts. The asbestos fibers can be detected,

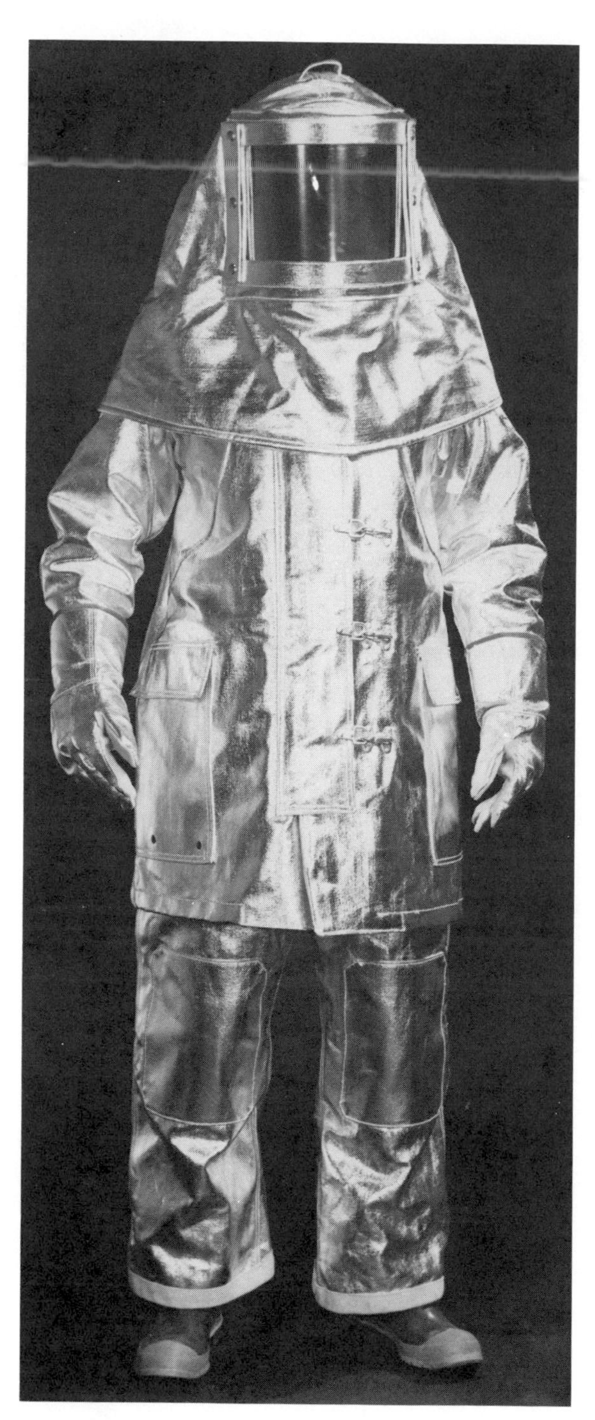

Fig. 5-1. Suits made from asbestos fibers are used by firefighters in extinguishing fires. This suit has been metal-coated for additional protection. (Courtesy Fyrepel Products, Inc., Newark, Ohio.)

but an accurate quantitative count usually is not possible. Some soils near asbestos mines contain considerable quantities of the mineral. Asbestos dust is carried great distances from mining and milling operations. In South Africa it was found that much of the airborne asbestos particles were blown off dumps and roads made from the mine tailings.

Mineral asbestos rock is crushed, dried, and again crushed before being processed to separate the fibers. It goes through various fluffing, shaking, and aspirating processes, during which the fibers are cleaned and graded to market specifications. Large quantities of dust are generated during this milling process. The processed asbestos is then manufactured into more than 3000 different products. Approximately two thirds of the asbestos used in the United States is in the construction industry.

The air around construction sites is frequently contaminated when a fireproofing material containing 10% to 30% asbestos is sprayed onto girders and other structural components. This causes problems not only around the construction site but the microscopic asbestos fibers may be carried 25 to 50 miles by the air. Even heavily travelled city streets are potential sources of asbestos pollution. Abrasion of automobile brake linings and clutch facings releases thousands of microscopic asbestos particles into the air. Everyone comes into daily contact with objects containing asbestos: wallboard, roofing shingles, gaskets, automobile undercoating, asbestos textiles, asbestos paper products, insulating materials for pipes and ducts, asbestos cement, floor tile, etc. In addition, recent investigations have found asbestos fibers in beer, wines, soft drinks, and tap water. Samples of tap water from Ottawa, Montreal, and Toronto contained from 2 to 4.4 million fibers per liter. Asbestos fibers have recently been discovered in injectable drugs. This contamination results from the use of asbestos filters in the preparation of the drugs. The irony of this is that the major reason for using asbestos was to filter out the contaminants, principally bacteria. In Japan, talc that is added to rice is frequently contaminated with asbestos, and this may be the cause of high rates of stomach cancer found there.

Preliminary measurements around several American cities showed that each cubic meter of air contained from 0.01 to 0.1 μg of asbestos. When one considers that city air usually contains about 75 to 200 μg of particulate matter, this seems like a very small amount. However, the asbestos fragments are so small that 0.1 μg may represent 100 million tiny fibers, each one capable of doing damage to body tissue.

The problem with determining safe levels of airborne asbestos is that it is not known how much disease is related to a given level of asbestos in the air, but it is now clear that exposure to asbestos is a health hazard. It causes or leads to asbestosis and lung cancer as well as pleural calcification, pleural plaques, and mesotheliomas. The problem is further complicated in that asbestos affects various people differently rather than on a linear dose scale; that is, some people seem to be much more resistant to its effects than are others.

Fate of inhaled fibers. Not all asbestos fibers that are breathed in actually reach the lungs. Some are stopped by the nasal hairs or the small size of the bronchioles, but those less than 200 μm long and 3.5 μm in diameter can enter the lungs. The fate of an asbestos fiber once inhaled and deposited in the lung is not fully understood. First of all, those fibers shorter than 0.5 μm have largely been ignored because they are difficult to study. Electron microscopic studies have shown that for every asbestos fiber recognizable under an ordinary microscope there are hundreds of tiny unrecognized particles that are just as dangerous as the longer fibers. Once in the lung, macrophage cells (Chapter 3) pick up the fibers by phagocytosis. The longer fibers may then be coated with an iron-containing compound (probably ferritin), and this structure is known as an "asbestos body." Formation of asbestos bodies seems to be a continuous process, with old bodies growing in thickness and uncoated fibers being converted into asbestos bodies months or years after entering the body. Many recent studies show that one fourth to one half of the urban population have these structures in their lungs. These are being inhaled either from the city air or from exposure to asbestos.

Extended exposure to asbestos results in asbestosis, causing a diffuse fibrosis in the lower lobes of the lungs. It is common among asbestos workers, although it usually does not develop until 20 to 40 years after the beginning of exposure. Because the latent period is so long, it is difficult to establish safe limits of exposure. In a survey of asbestos workers in the New York–New Jersey area 339 out of 392 workers with more than 20 years of exposure showed evidence of asbestosis. Cancer of the lung is also associated with exposure to asbestos. Some experts believe that the probability of cancer is proportional to the number of asbestos fibers, number of susceptible cells, concentration of contaminating substances carried on the asbestos fibers, and time of exposure. Once the asbestos fibers enter the body, the potential for cancer continues throughout life.

Asbestos fibers do not always remain in the lungs but may be transported to other parts of the body by macrophage cells in the lymphatic or venous systems. Asbestos bodies may be transported to the spleen, small intestine, and other areas. They may stay in the abdominal lymph nodes or in the mesothelium (the thin tissue covering the internal organs and lining the body cavities). These tumors of the mesothelium (mesotheliomas) are strongly associated with asbestos contamination. In one study of eighty-three patients with a diagnosis of mesothelioma 52.6% of the patients had a history of occupational or domestic exposure (living in a house with an asbestos worker). Among those with no occupational or domestic exposure to asbestos 30.6% lived within ½ mile of an asbestos factory. So far, there have been few cases of mesothelioma in the general population, but it is considered a frequent cause of death among asbestos workers.

Mechanism of action. The mechanism by which asbestos causes pulmonary fibrosis, cancers, and tumors is not understood, but several hypotheses have been advanced. One hypothesis holds that the fibers act as physical irritants, and the constant irritation induces a tumor after 20 to 30 years (the usual latent period). A second hypothesis is that the asbestos fibers contain small amounts of cancer-causing substances. Various contaminating carcinogens can even be picked up by the airborne fibers. There is evidence that the carcinogenic effects of asbestos can be accelerated by other pollutants, such as tobacco smoke. Some experts think that the asbestos itself does not cause the damage but that it is the vehicle which carries the carcinogens into the tissues. Nickel, chromium, and benzo[*a*]pyrene, which are all carcinogenic, are carried into the lungs and other organs by the asbestos. Once in the lungs, these chemicals can be dissolved in the plasma. As a third hypothesis, it has been proposed that the asbestos fibers accumulate in the lungs and form asbestos bodies (fiber covered by iron-containing material). After 20 years or more the asbestos bodies disintegrate, leaving the free asbestos fibers to cause asbestosis or cancer.

One of the most important proteins in animal cells is the fibrous material, collagen. Collagen fibers provide rigidity and strength to tissues. Ligaments and tendons derive their toughness from them. When asbestos dust was injected into rats, it stimulated the synthesis of collagen. This could be the mechanism by which asbestos causes asbestosis (fibrosis of lung tissue).

Perspective. Perhaps the greatest problem with asbestos is that people do not recognize its potential toxicity and work with it as they do with other harmless materials such as plywood or cloth. This casual attitude probably stems from the long latency period between exposure to asbestos dust and the first appearance of ill effects—usually 20 years or more. By the time the symptoms appear the individual may have no association with asbestos and probably does not associate the illness with the previous exposure.

Asbestos is commercially valuable and essential to our present life style. No one advocates banning the use of asbestos, but it is important that the public be made aware of the fact that asbestos can be a lethal material if handled improperly. Fortunately this is being done. For example, the Illinois Pollution Control Board has adopted a ban on the spraying of asbestos material and has tightened controls on plant emissions into the air and water. These regulations limit plants manufacturing asbestos products to no more than three asbestos fibers per cubic centimeter of air (that is still 2.295 million fibers per cubic yard). The Environmental Protection Agency is currently preparing emission standards that will apply to all mining, milling, spraying, and manufacturing of asbestos.

Establishing an allowable safe level for asbestos is a difficult task. One suggestion has been to use a time-weighted average concentration to measure the total dose. Multiplying the average number of fibers in the air by the years of exposure (based on an 8-hour day, 40-hour week) could give a rough guide to the total dose. Using this method, most workers begin to show evidence of asbestosis after exposure to 50 to 60 million particles per cubic foot-years. That is, if the air in a factory contained 5 million particles per cubic foot, the worker would probably develop asbestosis after 10 to 12 years of exposure.

Through the use of elaborate ventilating systems, filters, and carrying out some operations as wet processes to keep down dust, the pollution from factories has been greatly reduced. However, the workman who saws asbestos boards or asbestos cement pipes, strips insulation from wires, or installs insulation breathes these fibers in unfiltered air at construction sites every day. The razing of old buildings is another significant source of occupational exposure to asbestos-containing dust. When fireproofing asbestos is sprayed on steel girders in high-rise buildings, not only are the workers exposed to the spray, but other nonworkers are exposed as well, since construction of this type usually occurs in highly populated places.

It has been suggested that asbestos be treated like poison—avoid it if at all possible.

PHOSPHORUS

Phosphorus is abundant in phosphate rock and is essential to all forms of plant and animal life. It is present in the food we eat and the beverages we drink. In fact, it is found just about everywhere, including the atmosphere. Phosphates have been used in large quantities for many years and have never been known to create a health or safety problem for people. Why then has there been such a controversy in recent years over phosphate in detergents? The problem is not that phosphate (PO_4^{-3}) is bad, but

that it is *too* good. It causes things (particularly algae) to grow when no one wants them to grow. Lakes generally come into being as oligotrophic (little food) waters of low productivity, and as nutrients are continually added from the surrounding watershed, progress to a eutrophic (true food) condition with high nutrient levels and high productivity. This process is called *eutrophication*.

Nutrient enrichment of such waters causes increased growth of algae and other aquatic plants, deterioration of fisheries, and deterioration of water quality. But is eutrophication good or bad? The answer depends on the intended use of the lake. A beautiful lake with clear blue water is fine for domestic water supplies or boating and swimming. The water in these oligotrophic lakes is clear because there are few nutrients, which means few plankton, which means few fish. When nutrient material is regularly carried into the lake, plankton become plentiful and provide a good food supply for the fish—but the lake is no longer crystal clear.

Lakes and ponds become more and more eutrophic with age even in the absence of man, but man's activities can vastly accelerate the process. In many instances increasing concentrations of a number of nutrient materials has caused an excess growth of algae, which on decomposition depletes the oxygen supply and causes extensive fish kills. At this stage the commercial and sport fisheries begin to disappear. Only the hardy "trash" fish (catfish, carp, buffalo, etc.) can survive. It has been estimated that the maturation of Lake Erie, thought to be about 10,000 years old, has been advanced 150,000 years in the last century by man-caused pollution. The detergent industry has attempted to convince the public that since eutrophication is a natural process, accelerating it does not constitute water pollution. However, ecologists believe that any material that speeds deterioration of the environment is a pollutant. In this case the nontoxic normally beneficial phosphate must be considered such a material.

Cellular uses of phosphorus. Several cellular roles of phosphorus were described in Chapter 2, but the cellular mechanisms in which phosphorus is involved will now be reviewed briefly. Most cellular phosphorus occurs in the structure of DNA (chromosomes), RNA (protein synthesis), and phospholipids (structure of the membranes of the cell). Phosphorus also plays an important role (as a major component of ATP) in the energy transfer processes of the cell. If the supply of phosphorus is insufficient, new genetic material cannot be made (hence cells cannot duplicate), and the various membranes forming the cell's organelles cannot be formed (materially slowing cellular repair). If there is a surplus of phosphorus, rapid growth is possible, assuming that other essential materials are also available.

Phosphorus as a limiting factor. Plants require many elements for growth. Chief among them are carbon, hydrogen, oxygen, nitrogen, and phosphorus. The material that is in shortest supply (not in absolute amount, but relative to the amount needed) is known as the *limiting factor,* or the factor that by itself has the greatest effect on the growth of an organism. Chapter 1 described how all materials go through cycles. Of the five chief elements needed by living organisms, four pass through a gaseous phase somewhere in their cycle: carbon cycle, carbon dioxide; hydrogen cycle, hydrogen gas or water vapor; oxygen cycle, oxygen gas or water vapor; and nitrogen cycle, nitrogen gas or ammonia. Because of their large gaseous reservoirs, these four major nutrients are usually present in abundance in aquatic ecosystems. Phosphorus does not have a gaseous phase, and this means that it enters the environment only as a solid or dissolved in water.

The main reserves of phosphorus are phosphate rock or other phosphate deposits, and it is released by weathering, leaching, or mining. It moves through terrestrial and aquatic ecosystems by passing through the food chain and is released by death and decay (Fig. 1-10). However, much of it is lost to the sea, where it may again be taken up in a food chain by the plankton.

As members of the food chain die, the phosphorus that has been incorporated into their body tissues is carried to the bottom. Some of it may be carried to the top again by upwelling currents and again may enter the food chain. The activity of plankton is insufficient to keep the phosphorus in circulation, and more is lost to the bottom sediments than is being added from the weathering phosphate rock. The deposited phosphorus may eventually be lifted above the surface by geological activity, which will provide more phosphate rocks to start the weathering process and the phosphorus cycle all over again.

The limiting factor in eutrophication of a lake is not always phosphorus. Algae require fifteen to twenty different nutrients, and the one that runs out first will be the limiting factor. In most oligotrophic lakes phosphorus is usually the limiting factor, but it may be present in excess in nutrient-rich lakes. Relative to the Great Lakes, experts believe that phosphorus is still the limiting nutrient in Lakes Superior, Huron, and Michigan, but not in Lakes Erie and Ontario. It is probable that removing phosphorus from the waste water entering the lakes could again make it limiting in Erie and Ontario.

In extremely eutrophic soft-water lakes carbon may become the critical element because of the interdependence of algae and decomposing bacteria. Each day carbon dioxide is removed by the algae by late afternoon and regenerated during the night by bacteria. However, situations such as this are probably rare. It does not apply to Lake Erie, for example, because there are tremendous amounts of available carbon provided by the dissolved carbonates (CO_3^{-2}) in the lake.

Nitrogen is probably the limiting factor in most coastal and estuarine waters (such as Long Island Sound, San Francisco Bay, or Chesapeake Bay) and some lakes because of the overabundance of other nutrients. Although algal blooms on lakes have now become a common occurrence, usually the limiting factor simply is not known. Eutrophication resulting from phosphorus is largely a freshwater problem restricted to lakes or other shallow, slow-moving bodies of water. To illustrate, the mean depth of Lake Erie is only 21 meters (maximum = 60 meters), whereas that of Lake Ontario is 91 meters (maximum = 225 meters), and although Lake Ontario is downstream from Lake Erie, it is much less eutrophic.

Sources of phosphorus pollution. Most uncontaminated lakes contain 0.01 to 0.03 ppm of phosphorus. Below 0.01 ppm algal growth is severely limited, but at concentrations of 0.05 ppm profuse growth may occur. Thus a very small increase in concentration can produce dramatic changes in a lake. Although almost all of the publicity concerning the phosphate controversy over the past few years has involved detergents, there are many other sources of phosphates (Fig. 5-2). Present-day domestic waste water contains about 10 ppm of phosphorus, and about one half to two thirds of this is from phosphate detergents. The remaining one third to one half is from human and animal waste. Removing phosphates

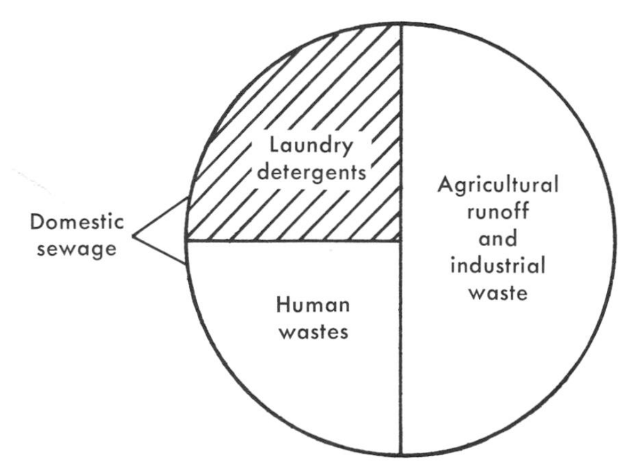

Fig. 5-2. Sources of phosphorus entering surface waters. Of all the phosphate in domestic sewage, about 1/2 to 2/3 comes from phosphate laundry detergents. Much of the phosphate in agricultural runoff is from phosphate fertilizers.

from detergents, then, will cut the phosphorus content of municipal waste water by 50%. This decrease in phosphate will certainly slow down the eutrophication process in many bodies of water. The phosphorus contribution from human waste and agricultural runoff (animal waste and fertilizers) would still be present, but a giant first step toward control of man-speeded eutrophication would have been taken.

Why do detergents contain phosphates? Although detergents have been available since the 1930's, they did not become popular until 1947—the year the phosphates were added. The basic ingredient of any detergent is the surfactant, a type of wetting agent that is responsible for the actual soil removal. One widely used surfactant is linear alkyl sulfonate (LAS). Unfortunately, surfactants do not work well in water containing calcium or magnesium ions (hard water). These ions interfere with the surfactant, and consequently some of the dirt remains in the clothes. This problem is remedied by adding a "builder" to the detergent. The builder, most commonly a phosphate, binds up the calcium and magnesium and also provides the alkalinity that is essential for effective soil removal.

In most detergents the phosphate builder accounts for 35% to 55% of the total weight. Nonphosphate detergents substitute another compound, such as sodium nitrilotriacetate (NTA) for the builder or use a surfactant that works without a builder. NTA is about as effective as phosphate in detergents, but there is still some question about its safety when large amounts of it are used. The other nonphosphate detergents currently available use precipitating builders that combine with the calcium and magnesium ions and form insoluble precipitates. These builders (silicates and carbonates) can cause a residue buildup in cloth or washing machines.

Of lesser importance, but still a problem, is phosphorus air pollution. The major sources of this pollution are the industries engaged in the production of phosphate fertilizers, phosphoric acid, and other chemicals for industrial use. Phosphate rock dust may be emitted in significant quantities from fertilizer manufacturing plants. There is a possibility that automobile emissions may contain toxic phosphorus compounds. Organophosphorus pesticides (Chapter 7) may also result in environmental pollution.

Perspective. Fortunately, phosphates are not a serious problem in rapidly flowing rivers or in coastal areas. Algae do not grow well in fast-moving streams, and phosphate is not a limiting factor in most bodies of salt water. Also, those homes that discharge their waste into septic tanks do not add to the eutrophication of local waters, since phosphorus is quite immobile in undisturbed soil. It is estimated that 80% to 85% of all the homes in this country can continue to discharge phosphates into the sewage systems without doing serious harm, although the streams would certainly be better off if the phosphates and other nutrients were not there. The remaining 15% to 20% of the households and industries have a very serious problem. Where are these problem areas? Any person living in a city whose sewage empties into (1) a lake, (2) a river, (3) a stream that enters a lake, or (4) a slow-moving canal is contributing to the phosphate problem. Some of the critical problem areas are the Great Lakes, southern Florida, lakes of the Tennessee Valley Authority (TVA), and Lake Tahoe.

Most present-day sewage plants have primary and secondary stages of treatment, and these are designed to produce an effluent rich in phosphates and other nutrients. Until recently, the only practical way of removing these nutrients from the waste water before it reentered the lake or stream was through a complex and expensive tertiary treatment. Now removal can be accomplished by various types of simple chemical treatment at relatively low costs. Good results can be obtained by adding either aluminum sulfate or ferric chloride to the aeration chamber at the end of the secondary stage.

For just over a penny per day per person the phosphorus level in the effluent water can be reduced

Fig. 5-3. For legend see opposite page.

to about 0.2 ppm. Since this effluent is diluted as it enters the stream or lake, the streams could be purer than they have been since we began using indoor plumbing. In the long run (15 to 20 years, at least) phosphorus removal from municipal waste streams seems to be a feasible goal. For the immediate future, however, removing phosphates from detergents will be more likely. Several states and many smaller governmental units have enacted laws to accomplish this goal, and it seems likely that others will soon follow.

NITRATES

A few plants such as clover can supply their nitrogen needs from the air, but for most plants the nitrogen in the atmosphere is an "unavailable pool." As plants grow, they use nitrogen from the soil, and this lost nitrogen must be replaced if soil fertility is to be maintained. Nitrogen is regularly added to the soil from the atmosphere by lightning or by the nitrogen-fixing bacteria in the soil (Fig. 1-9), but the rate is too slow to replace the nitrogen lost as a result of agriculture. Organic fertilizers (manure, blood, bones, guano, etc.) have been used for centuries, but a decreasing supply of organic fertilizers and an increasing population caused a shift from organic fertilizers to chemical ones (Fig. 5-3). The synthesis of ammonia (NH_3) from nitrogen (N) and hydrogen (H) in 1910 was an important development that made chemical fertilizers economically feasible. Today farmers can buy chemical fertilizers containing almost any combination of plant nutrients. Most agricultural ammonia is converted into solid derivatives such as ammonium nitrate (NH_4NO_3) or ammonium chloride (NH_4Cl). The U. S. Department of Agriculture has estimated that the use of chemical nitrogen fertilizers will be increased about ten times between 1970 and 2000.

Nitrogen compounds in the ecosystem. As dead plant and animal bodies are broken down by bacteria and fungi, the protein in their tissues is transformed into ammonia. Another group of bacteria (nitrite bacteria) degrade the ammonia to nitrites (NO_2^-), and these nitrites may then be transformed into nitrates (NO_3^-) by the nitrate bacteria as follows:

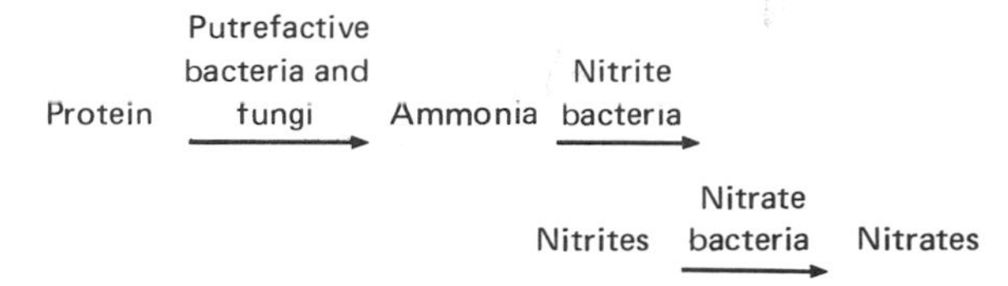

Nitrate nitrogen is quickly available to root systems, where it can be absorbed and built into new protein. However, nitrates can be leached from the soil by flowing water.

As a result of leaching and other losses, rarely is more than 75% of the fertilizer applied to a field taken up by the agricultural crop: the other 25% finds its way into ground and surface waters. The magnitude of this problem becomes apparent when one considers the amount of fertilizer used on crops. In Iowa, farmers used 692,379 tons of nitrogen (as N) in the 1970-1971 fertilizer year. Although the nitrate

Fig. 5-3. Nitrate is an important fertilizer added to crops. **A,** Animal manure has been used for centuries as the primary form of nitrogen fertilizer. **B,** In recent times synthetic nitrogen fertilizers in the form of ammonia (here) or ammonium nitrate have been used extensively because of their ease of application and the limited supply of animal manure. (Photographs by David Osgood.)

level in the Mississippi River seems to have stabilized since 1960, records from sixteen smaller streams show that over a 5-year period the averages for these streams were as follows:

1962	1.8 mg per liter (as nitrate nitrogen)
1963	1.7 mg per liter
1964	1.8 mg per liter
1965	1.9 mg per liter
1966	2.2 mg per liter

Although the nitrate levels in most midwestern streams are not alarming at this time, the concentrations do appear to be increasing. The U. S. Public Health Service allows no more than 10 mg per liter of nitrate nitrogen (or 45 ppm of nitrates) in drinking water. What would the excess nitrogen do in the ecosystem? First the effect on the soil itself will be examined.

Most good soils include a large amount of organic matter called humus. Most of the natural nitrogen in the soil is in complex molecules in the humus and is derived from decayed woody fibers, animal manure, and other decaying substances. These materials are important in maintaining the texture of the soil. Inorganic nitrogen (the type used in chemical fertilizers) amounts to less than 1% of the total soil nitrogen. When only inorganic fertilizers are applied to the soil, the crop yield is high, but often there is a decrease in the amount of humus. The soil can become compacted, and this in turn leads to loss of nitrates from the soil. The continued use of chemical fertilizers does not replenish the humus and may lead to an increase in the amount of nitrogen lost through runoff. Although there has been a 50% reduction in the original organic nitrogen content of the midwestern soils, they are still well supplied with organic humus. However, the soils of the northeastern United States and central Europe need great amounts of organic humus to replace that lost through years of leaching. With the continued loss of humus from the soil the efficiency of nitrate transfer from soil to plant decreases, necessitating the use of increased amounts of chemical fertilizers to maintain crop yields.

The nitrogen that is lost from the fields will be considered now. Some of the nitrogen lost from the agricultural fields is converted into atmospheric nitrogen by the denitrifying bacteria. This is an economic loss but is of no environmental consequence. The rest of the nitrogen that is lost either enters the ground water or surface waters. On entering a lake, the nitrogen fertilizes the aquatic plants just as it did the agricultural crops. This addition of plant nutrients can lead to excessive growth of aquatic weeds and algae. This eutrophication process occurs in any body of water that is oversupplied with nutrients. As explained in the section on phosphates, the nutrient that is in shortest supply will be the one that limits noxious plant growth. In most freshwater lakes and ponds this limiting nutrient is phosphate, not nitrates; but in coastal marine waters, nitrogen is the critical limiting factor to algal growth and eutrophication. One reason why phosphate is less critical in coastal marine waters is that phosphorus is regenerated more quickly than are nitrogen compounds from the decomposing organic matter. It is possible that replacing the phosphates in detergents with a nitrogen compound (NTA) will worsen the pollution of saltwater bays near large cities.

Dietary intake of nitrate. Most of our dietary nitrate intake is from vegetables or water supplies that are high in nitrate content and from the nitrates used in curing meats. Many vegetables contain very high concentrations of nitrates (3000 ppm or more). Although there is considerable variation due to age, environment, or particular variety, nitrate levels are generally high in spinach, turnip greens, celery, lettuce, radishes, beets, and eggplant. Some of the factors that tend to increase nitrate content are (1) high levels of fertilization, (2) lack of water, (3) plant damage from chemical treatments, (4) deficiency of some nutrient, or (5) reduced light during maturation.

Nitrates and nitrites are used extensively for preservation of meat products and for color fixation in foods. In meat processing, nitrite is added to give the characteristic red-pink color to cured meat and to produce the cured meat flavor. Because of its antibacterial properties, nitrite is used to preserve canned meats (particularly smoked fish) processed at less than sterilizing temperatures. The nitrite prevents the growth of bacteria that cause botulism, a disease fatal to man. Apparently the nitrite stops cell division in these cells. Because nitrites are hazardous to health and because of the danger of botulism, it has been necessary to set both maximum and minimum levels for nitrites in processed fish. For example, smoked chub must contain not less than 100 mg of nitrite per kilogram of fish; salmon must not contain more than 200 mg per kilogram. Although the 200 mg per kilogram limit (200 ppm) has been established as the legal limit for meats, the daily intake for man should be limited to 0.4 mg per kilogram of body weight. (This is about 0.001 ounce for a 150-pound person.)

Although it is not a general problem at this time, the water supplies of several agricultural states in the United States are loaded with nitrates (e.g., the central valley of California and the agricultural states of the Midwest). In one study in Illinois it was found that at least 55% to 60% of the nitrates in the surface waters entering Lake Decatur were from nitrogen fertilizers. The U. S. Public Health Service acceptable standard for drinking water is not more than 45 mg of nitrate per liter of water (45 ppm), but concentrations in many Illinois streams and water supply reservoirs have already equaled or exceeded this standard. In addition, in some areas nitrates in well water are found at naturally high levels.

Nitrates are not especially dangerous by themselves, but under certain circumstances they can be the starting point for a series of chemical reactions that converts them into toxic substances. At low concentrations the nitrates are excreted in the urine, but when the total intake is increased, bacteria may convert the *nitrate* ions (NO_3^-) into highly toxic nitrite ions (NO_2^-). Nitrate is reduced by bacteria (particularly the coliform group, or *Clostridium*) only when growing under anaerobic (without oxygen) or partially anaerobic conditions. These microorganisms substitute nitrate for oxygen as a final hydrogen acceptor (Chapter 2) in the energy-releasing mechanism as follows:

$$NO_3^- + NADPH_2 \longrightarrow NO_2^- + NADP^+ + H_2O$$

Since it is only the nitrites that are hazardous, the nitrates are a problem only when conditions exist that permit large amounts of nitrite to be produced. Bacterial action in damp animal feeds with a high nitrate content can produce sufficient nitrites to be toxic to livestock. The rumen of cattle and large intestine of horses both provide an ideal environment for bacterial growth, and this poses a real danger whenever the food or water of farm animals is high in nitrate. Except for extreme cases, nitrates are little problem for adult humans, but they do constitute a hazard for babies under 4 months old because of the lower acidity of their stomachs. This permits bacteria native to the alkaline large intestine to survive in the upper part of the digestive system. There are many instances of infants being poisoned by eating spinach that was not refrigerated after the jar of baby food was opened. The large amount of nitrates in stored spinach (raw or cooked) can be reduced to nitrites. Any food with a high nitrate content can be a hazard.

Mechanism of action. The nitrites by themselves can cause detrimental effects, but they can also react with other substances called amines to form new toxic compounds. Nitrites themselves will be considered first.

In normal situations hemoglobin in the red blood cells carries oxygen (as oxyhemoglobin) to all the cells in the body. When high levels of nitrite are present, the nitrite binds itself to the hemoglobin (forming *methemoglobin*) more readily than does the oxygen. Since more nitrite is present, more methemoglobin is formed, thereby causing a depletion of oxygen in the body tissues. Only in extreme cases does this affect adult humans, but the gastrointestinal tracts of infants and farm animals are much more likely to have appropriate conditions for conversion of large amounts of nitrate to nitrite. Since the methemoglobin does not carry oxygen, labored breathing can progress to death from suffocation. Drinking water containing nitrate concentrations greater than 45 ppm can cause methemoglobinemia in infants, although most cases have involved NO_3^- concentrations in the 66 to 1100 ppm range.

Nitrites may also be converted in the body to cancer-causing substances called *nitrosamines.* The carcinogenic nature of a wide variety of nitrosamines has been proved. They can also produce mutations. It is possible that nitrosamines are formed any time their components (nitrites and amines) are found together in the environment. It is known that certain foods contain nitrosamines, but they can be formed in the stomach as well. This can be illustrated by the following example. One of the major constituents of meat is creatine. Under acid conditions, such as in the stomach, nitrite becomes nitrous acid (HNO_2). Creatine and nitrous acid can undergo the following reactions

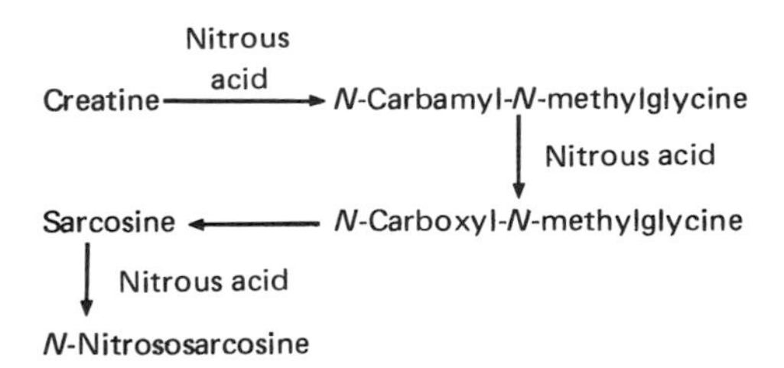

to form *N*-nitrososarcosine, which has been shown to induce cancer of the esophagus in rats. Whether reactions such as this actually occur in the human stomach or whether they induce human cancer is not known for sure. Since nitrosamine compounds are used in a number of industrial products, they should be handled cautiously, since they can be picked up by inhalation, topical application, or orally.

Metabolic products of nitrosamines can attack guanine in nucleic acids (Chapter 2), which can cause mutations in DNA or RNA. Again, this information is based on animal studies, and its applicability to man is not established.

Perspective. The need for nitrates in fertilizers is great, but the potential hazard of nitrates, nitrites, and nitrosamines resulting from their use must be recognized and dealt with. If increasing food production can be maintained only by increasing the use of fertilizers, then some way must be found to keep the fertilizers on the fields and out of the water supplies. Various changes in agricultural practices could reduce the problem. Since no-tillage (nonplowing) crop production eliminates soil movement, this would help to stop fertilizers and pesticides from being carried off by soil erosion. If the underground water supplies become contaminated with nitrates, it will take an extremely long time for them to be cleansed of their pollutant. There is also a problem with nitrite food preservatives. If use of the preservative is limited, far worse situations may occur from improperly stored food. If reasonable care is taken in the use of all these substances, most serious problems could be avoided. The effects of increased nitrogen in water supplies should certainly be considered before a nitrogen-containing substance like NTA is used as a phosphate substitute in detergents.

CARBON DIOXIDE

Every student of science, biology, or chemistry has seen the following equation:

$$\left.\begin{array}{l}\text{Sugar}\\\text{Coal}\\\text{Oil}\\\text{etc.}\end{array}\right\} + O_2 \longrightarrow CO_2 + H_2O + \text{Energy}$$

As foods and fossil fuels (oil, coal, gas, etc.) are burned, complex organic molecules break down into the simpler molecules, carbon dioxide and water. The production of energy from fossil fuels represents a great amount of combustion. Most of the coal is used for generation of electricity, and the natural gas and oil are used for heating purposes and as fuel for vehicles. The increased worldwide use of these fuels in the last few decades has been spectacular:

COAL (metric tons)		OIL (barrels)	
1870	250,000,000	1890	Negligible
1970	28,000,000,000	1970	12,000,000,000

Presently the use of oil is increasing by 7% each year, which means that the demand for oil is doubling every 10 years. One of the greatest effects of the increased use of fuels is the increased emission of carbon dioxide. Fig. 5-4 shows that the concentration of CO_2 in the atmosphere has increased from about 290 ppm in 1860 to 321 ppm in 1970. Since 1950 the CO_2 concentration has increased at an annual rate of 0.2%.

Increasing the atmospheric concentration of CO_2 is known to stimulate more rapid growth in plants, and this effect has been capitalized on by greenhouse owners. Much discussion has been devoted to the possible effects of this increased concentration of CO_2 on the world's climate. When shortwave radiation from the sun strikes the earth, it is transformed into long-wave radiation and is re-radiated from the earth as heat. Normally the amount of energy (heat) reaching the earth is balanced by the amount leaving, and the mean temperature of the earth remains relatively constant so that the climate in a given place does not change. Short-term geographical variations in this heat balance result in summers and winters. Carbon dioxide molecules have a strong absorption band at wavelengths between 1500 and 1800 nm (Fig. 1-2), and this is the region of the spectrum where most of the energy radiating from the earth is concentrated. Based on this, various scientists have proposed that the increase in CO_2 could increase the temperature of the earth by several degrees. However, recent research shows that as more CO_2 is added to the atmosphere, the rate of temperature increase becomes proportionally less and less and eventually levels off. As more and more CO_2 is added to the atmosphere, the CO_2 absorption band "saturates." Further increases in CO_2 do not cause corresponding increases in long-wave (infrared) absorption so that further increases in CO_2 are unlikely to cause a significant temperature change. The concentration of CO_2 in the atmosphere has increased about 10% in the last century, and calculations show that an additional 10% increase (up to about 350 ppm) might raise the temperature no more than 0.2° F. Even if a CO_2 increase were to occur, the higher levels probably would not persist for long. The oceans contain about sixty times more carbon dioxide (much of it in the form of carbonates, CO_3^{-2}) than the atmosphere and will absorb much of the excess.

Although current theories indicate that CO_2 resulting from combustion of fossil fuels will not affect the global temperature, another combustion waste product, particulate matter or aerosols, may cause critical temperature changes. Depending on the composition, number, size, and shape of the particles, the suspended matter in the atmosphere could cause the

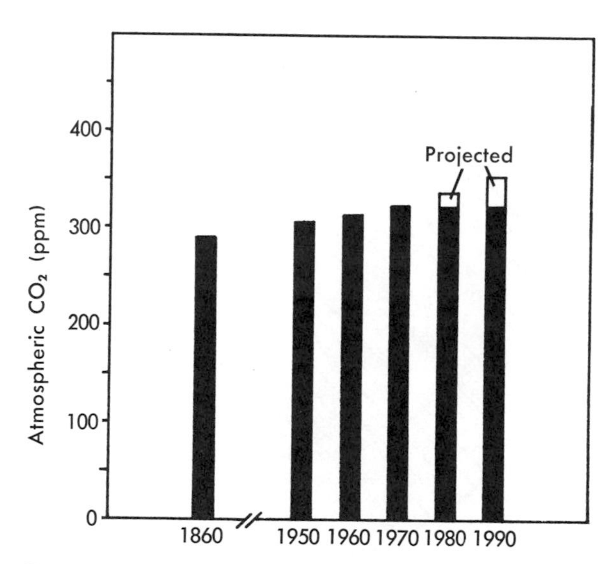

Fig. 5-4. The change in carbon dioxide concentration in the atmosphere in the last 100 years. This change is believed to be due to the increased burning of fossil fuels and greater agricultural activities. (Data from various sources, including Man's impact on the global environment: report of the Study of Critical Environmental Problems [SCEP], Cambridge, Mass., 1970, The MIT Press.)

earth's temperature to increase or decrease. On the one hand, it could increase absorption of solar (and other) radiation, which would tend to warm the earth. Conversely, it could increase the reflectivity of the earth's atmosphere and reduce the amount of incoming radiation, which would result in the cooling of the earth. No one knows which, if either, will happen.

CARBON MONOXIDE

Carbon monoxide (CO) is an odorless, colorless gas that is found in high concentrations in urban atmospheres. In fact, no other gaseous air pollutant with such a toxic potential as CO exists at such high concentrations in urban environments (Fig. 5-5). Since it has no warning properties such as odor or color, an individual is usually aware of its presence only after symptoms such as headache, dizziness, nausea, or difficulty in breathing appear. Contrary to popular belief, fires in a closed space are dangerous not because they exhaust the oxygen but because they emit CO. House fires produce CO concentrations up to 50,000 ppm.

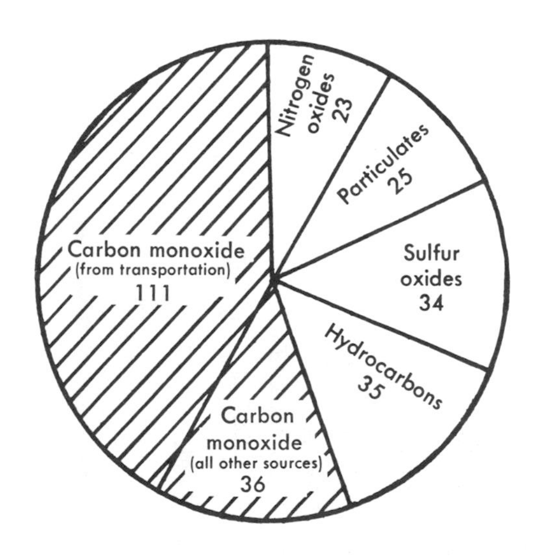

Fig. 5-5. Total weight of United States air pollutants in 1970 (units = millions of tons). (From U. S. Environmental Protection Agency: Nationwide air pollution emission data, 1940-1970, Washington, D.C., 1972, U. S. Government Printing Office.)

Surprisingly, natural processes produce at least ten times more CO than all automotive and industrial sources combined. Most soils contain bacteria that can convert CO into CO_2. This means that the average Temperate Zone soil can convert over 200 tons of CO per square mile per year. Reactions in the atmosphere also convert large quantities of CO by oxidation to CO_2. Natural processes produce more than 3.5 billion tons annually worldwide, whereas man-made CO amounts to about 270 million tons —less than 10% of the total. Why, then, is CO a serious pollutant?

In urban areas almost all the CO comes from automobiles. For example, 97% of the CO in New York City is from internal combustion engines. Large industrial furnaces rarely emit CO in significant quantities when combustion is occurring at maximum efficiency. The carbon monoxide problem is much more likely to come from the totality of automotive and household sources in a community. A worker in an industrial plant is protected from hazardous CO levels by regulations that permit no more than 50 ppm of CO, but as he drives home from work he may be exposed to CO levels of 115 ppm in downtown traffic. Because so much CO can accumulate in such restricted places—in a car, building, or city street—it does pose a pollution problem. Although industrial standards are set at 50 ppm, it is known that exposure to CO levels as low as 10 to 15 ppm for 8 hours can produce adverse effects. Older research studies seemed to rule out the possibility of subacute or chronic effects of CO at levels much lower than those recognized as toxic, but recently more sophisticated techniques have led researchers to believe that chronic effects may result from long exposure to low levels of CO.

Mechanism of action. It is now believed that the

sole mechanism for all the physiological effects of CO is lowering the oxygen-carrying capacity of the blood, and all the various symptoms of CO poisoning are related to the resulting hypoxia or lack of oxygen. The affinity of hemoglobin (Hb) for CO is over 200 times greater than its affinity for oxygen, and a binding site on the hemoglobin molecule cannot be occupied by both CO and oxygen. Therefore, any hemoglobin molecule that is carrying carbon monoxide (HbCO) is not available for carrying oxygen (HbO_2). Earlier it was mentioned that hemoglobin normally carries oxygen to the body tissues and releases it (Chapter 3), but when hemoglobin binds with CO, the combination is more or less permanent. Carbon monoxide also reduces the dissociation capacity of oxyhemoglobin (HbO_2) so that even if the hemoglobin is carrying oxygen, it does not release it to the tissues. The amount of carboxyhemoglobin (HbCO) formed depends on the concentration of CO in the air and the duration of exposure.

Impairment of cognitive and psychomotor abilities occurs when as little as 5% of the hemoglobin is in the HbCO form. Recent research with rats shows that exposure to CO at 50 ppm (the accepted "no effect" level) for 5 hours a day 5 days a week for 12 weeks caused a significant loss of trace metals from the liver. This indicates an overall reduction in cellular respiration and ATP production. Hypoxia caused by CO inhibits the function of alveolar macrophages (Fig. 3-2), weakening tissue defenses against airborne bacterial infection. Maternal CO poisoning during pregnancy can cause fetal death, which is due primarily to lack of oxygen in the fetal circulatory system. Carbon monoxide poisoning causing unconsciousness for 30 minutes to 5 hours does not do permanent damage to the mother but can cause brain damage, idiocy, or death to the fetus. The severity of damage is related to the month of pregnancy, the fetus being particularly vulnerable shortly before birth.

In addition to its effects as a primary pollutant, CO also reacts with other materials in the atmosphere to form some of the components of photochemical smog, which is discussed later in this chapter. Briefly, CO greatly accelerates the conversion of nitric oxide to nitrogen dioxide, the appearance of ozone, and the disappearance of olefins. Carbon monoxide is also involved in a hydroperoxyl reaction in which CO is converted to CO_2. It is apparent that CO is not an inert gas.

Perspective and environmental standards. Fortunately, the historical trend indicates that carbon monoxide levels in cities are decreasing, as demonstrated by the following figures*:

New York City

YEAR	RANGE	AVERAGE
1932	2–129 ppm	32 ppm
1966	19– 95 ppm	32 ppm
1967	1– 17 ppm	8 ppm

Paris

YEAR	RANGE
1928	40-60 ppm
1959-64	Exceeded 10 ppm only 91 times in 6-year study

Much effort is being exerted to reduce the carbon monoxide output from automobiles. In California in 1962 a survey revealed that automobile exhaust was 3.2% carbon monoxide. The United States standard for 1970 was 1%, and the suggested objective for 1975 automobiles is 0.5% carbon monoxide. The ultimate objective is to reduce this to 0.25%. This is near the theoretical lower limits for automobile exhausts. The global atmospheric concentration is 0.00003%.

SULFUR DIOXIDE

In urban air, sulfur (S) is one of the most troublesome of air pollutants. The many compounds of sulfur enter the atmosphere from a variety of

*Data from Eisenbud, M., and Ehrlich, L. R.: Science **176**:193-194, 1972.

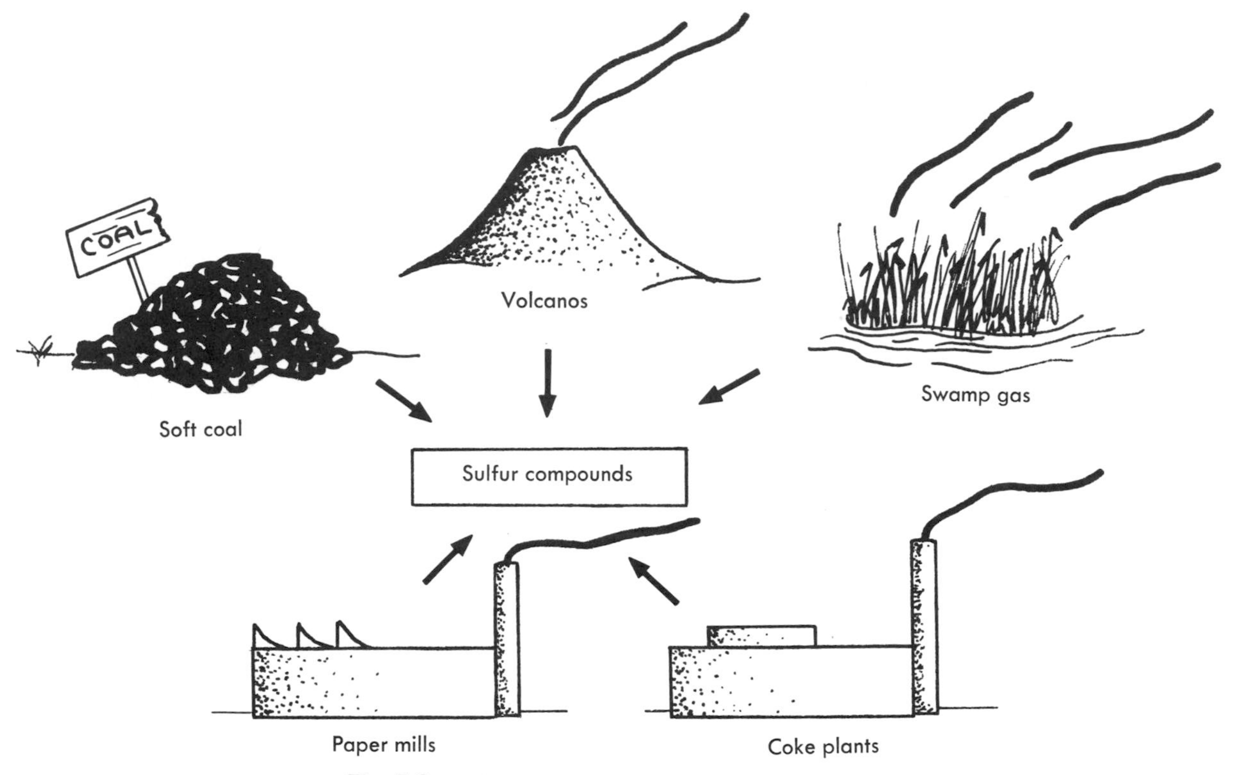

Fig. 5-6. Major sources of sulfur entering the atmosphere.

sources (Fig. 5-6). Long before man learned to use sulfur, the swamps, shallow lakes, and the ocean added large amounts of hydrogen sulfide (H_2S) to the atmosphere. Volcanic activity introduces hydrogen sulfide and other sulfur compounds into the air. Situations as varied as decaying meat, manufacture of kraft paper, and piles of coal mine refuse also contaminate the air. However, the major pollution problem comes from the burning of coal and oil, which contain sulfur as an impurity. When these fuels are burned, the sulfur also burns and enters the atmosphere. Man is now believed to be contributing slightly more sulfur to the atmosphere than such natural sources as volcanoes and the hydrogen sulfide derived from decaying organic matter. Of man's sulfur emission, about 95% is in the form of sulfur dioxide. Of the total sulfur dioxide emission, about 70% is from coal combustion (mostly for electrical power generation), 16% is from combustion of petroleum products, and the remaining emissions are from petroleum refining and nonferrous smelting. As with all matter, sulfur follows a cycle (Chapter 1) and does not remain in the air. It is eventually washed out in rain, enters the rivers, and finally ends up in the oceans. It is a problem to man while it is in the air, since it is washed out in rain, and to a lesser extent while it is in the rivers.

Reactions of sulfur compounds. The major sulfur compounds detected in the atmosphere are hydrogen sulfide (H_2S), sulfur dioxide (SO_2), sulfur trioxide

(SO_3), sulfuric acid (H_2SO_4), and sulfates (SO_4^{-2}). Hydrogen sulfide can undergo a number of reactions involving atomic oxygen (O), ordinary molecular oxygen (O_2), and ozone (O_3) to form mainly sulfur dioxide and water.

$$H_2S + \left\{ \begin{matrix} O \\ O_2 \\ O_3 \end{matrix} \right\} \longrightarrow SO_2 + H_2O$$

This reaction may be very fast in sunlight or when fog or cloud droplets are present in the air. Sulfur dioxide itself is not especially toxic, but in the atmosphere it can undergo further reactions to form sulfur trioxide (SO_3).

$$SO_2 + \left\{ \begin{matrix} \text{Sunlight} \\ O \\ O_2 \end{matrix} \right\} \longrightarrow SO_3$$

This almost immediately reacts with water to form sulfuric acid (H_2SO_4), a highly corrosive compound.

$$SO_3 + H_2O \longrightarrow H_2SO_4$$

The sulfuric acid is then either washed out by rain or converted to other sulfates (SO_4^{-2}), which settle out or are also washed out by rainfall. Most of these reactions occur much faster when the sulfur is adsorbed onto fine particles of fly ash in the air. Airborne water droplets containing metal salts that serve as catalysts greatly accelerate the reactions (Chapter 4).

Sulfur in the environment. The sulfur content of fuel oils varies from about 0.2% (some Algerian oil) to about 5% (some Mexican oil). Overall, the average for most coal and oil is about 3%. Burning of these fuels results in ambient urban air with SO_2 concentrations from a few hundredths of 1 ppm to a little over 1 ppm (rarely). Typically, these levels are below that necessary to produce demonstrable physiological or pathological damage. In unpolluted air such as might be found over the Atlantic Ocean, the SO_2 level is less than 0.01 ppm. In New York City in 1966 the average SO_2 concentration was 0.21 ppm, with peak concentrations of 2 ppm, but even this level is not usually perceptible to a normal person. Sulfur dioxide can be detected in the air at levels of 3 ppm and cause discomfort when the levels reach about 10 ppm, at which time eye, nose, and throat irritations begin to occur. The Environmental Protection Agency, however, has set much lower values as the maximum permissible levels: 0.03 ppm as the annual average and a maximum of 0.14 ppm on any single day. Why is there such a great difference between the maximum allowable limits and the level where people begin to feel its effects? Sulfur dioxide can cause considerable economic loss when it is converted into sulfuric acid. This acid attacks building materials such as limestone and mortar and causes rapid deterioration. It corrodes metal and damages paper and leather. Textile fibers (cotton, nylon, rayon, etc.) lose their tensile strength upon exposure to SO_2 or its by-products. In addition to prolonging the drying time of paints it makes them less durable.

The effects of SO_2 are not restricted to the air. Whether in fog, rain, or a lake, the SO_2 is converted to sulfuric acid. Over the past few years there has been a significant change in the acidity of many rivers, and this appears to be linked to the increasing levels of sulfur (and other) compounds. Pure rainwater (saturated with CO_2) will normally be slightly acid* (pH 5.7). However, measurements over a 3-year period in the northeast United States showed that the rain was much more acid (pH 4.4 average). This difference appears to be due to the sulfuric acid from the atmosphere. Although this does not pose any health hazard, the environmental consequences of this become obvious when one considers that certain aquatic animals (salmon, for instance) cannot survive when the pH of the water drops below 5.5.

Mechanism of action. Sulfur dioxide is classified as a mild respiratory irritant, and about 95% to 98% of

*Acidity is measured on a scale running from pH 1 to pH 14. The pH indicates the concentration of the hydrogen (H) ion. A solution at pH 1 is extremely acid, at pH 7 it is neutral, and at pH 14 it is extremely alkaline.

the gas is absorbed in the upper respiratory tract and never reaches the lungs. It penetrates into the lungs only when it is adsorbed onto the surface of particulate matter or converted to the sulfate form in an aerosol (fog, mist, metal vapor, etc.). These particles, usually less than 5 μm in diameter, get into the alveoli and have a far greater irritant potency than does SO_2 gas. Concentrations of about 1 ppm of pure SO_2 are required before any significant effects are expected, and 10 ppm will produce eye, nose, and throat irritation, especially in asthma patients. In fact, asthma patients are so sensitive that less than 10 ppm of pure SO_2 caused bronchoconstrictor symptoms, whereas normal patients showed almost no effect even in the presence of 30 ppm of SO_2. However, the proper way to assess the importance of SO_2 as an air pollutant is in terms of its ability to form irritant particles and not just its concentration in the gaseous state. The presence of aerosols or particulate matter capable of oxidizing SO_2 to sulfuric acid can cause a threefold to fourfold increase in irritant potential.

The absorptive capacity of the nasal cavity is very great. When rabbits are exposed to SO_2, an inhaled concentration of 200 ppm is reduced to about 10 ppm by the time the air reaches the trachea. The amount of SO_2 that reaches the lungs is so small that it usually does not cause a response at that site. It penetrates to the lungs only when adsorbed onto the surface of dust particles or when it is converted to the sulfate form in an aerosol. Experiments with dogs show that SO_2 is rapidly absorbed from the trachea (and lungs) and is distributed to all the tissues, including the brain.

Numerous experiments have been done to determine the effects of SO_2 and its by-products on animals. Monkeys were exposed to concentrations of 0.5 and 10 ppm of SO_2 for 22 hours a day for 2 years with no detectable changes in any physiological function or in any organ. In another study monkeys exposed to SO_2 at concentrations of 0.14 to 1.28 ppm showed no deleterious effects, but exposure to 4.69 ppm for 30 weeks did cause alterations in lung tissue. Guinea pigs exposed to 5.72 ppm of SO_2 for 12 months showed an increase in size of liver cells. Surprisingly, they had a *lower* incidence of spontaneous disease. Acute poisoning due to accidental overdose of SO_2 causes damage in nearly all organs and tissues, especially the lungs, heart, and brain. Workers in pulp mills exposed to SO_2 in concentrations varying between 2 and 36 ppm had a significantly higher frequency of cough, expectoration, and dyspnea (difficulty in breathing) on exertion than did nonexposed individuals.

There are conflicting opinions as to whether SO_2 slows down the beat of cilia lining the respiratory tract. If ciliary movement is stopped, it hinders the clearance of microorganisms and toxic substances from the respiratory tract (Chapter 3).

In addition to the local effects on the respiratory system, SO_2 inhalation can cause acidosis resulting from a decrease in the alkaline reserve in the blood. This causes disturbances in bone growth in children or degeneration of heart muscle in adults. Sulfur dioxide can be transformed in body fluids to sodium acid sulfate ($NaHSO_4$), which can cause chromosomal changes. Mutations have been produced in bacteria by exposure to SO_2.

Sulfur pollution also affects plants. It is absorbed through the breathing pores (or stomata) of leaves (Fig. 5-7). The initial injury usually appears in the spongy parenchyma cells. Later the palisade layer immediately above may be affected. These areas appear water-soaked. As the tissue dies, it becomes dry and papery and bleaches to a light tan color. The veins may remain green. Exposure to any given amount of SO_2 over a long term does less damage than exposure to the same amount in a short time. Apparently plant cells can detoxify a certain amount of SO_2 or sulfite (the product of SO_2 in solution), but when that level is exceeded, the water in the cell cannot be controlled and the cells collapse. When white pines were injured by SO_2 (0.06 ppm for 4 hours), the affected needle tissue collapsed, ac-

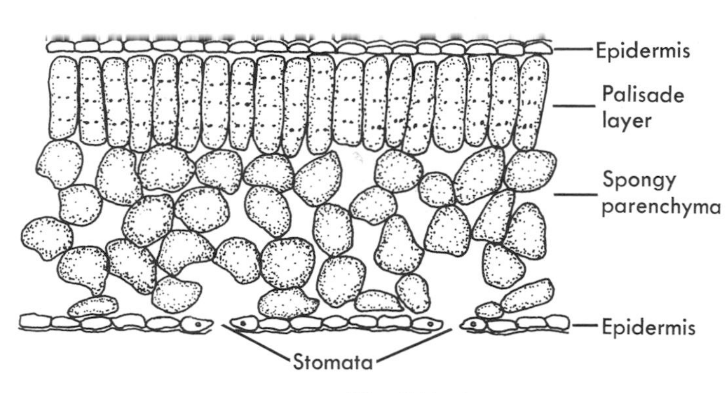

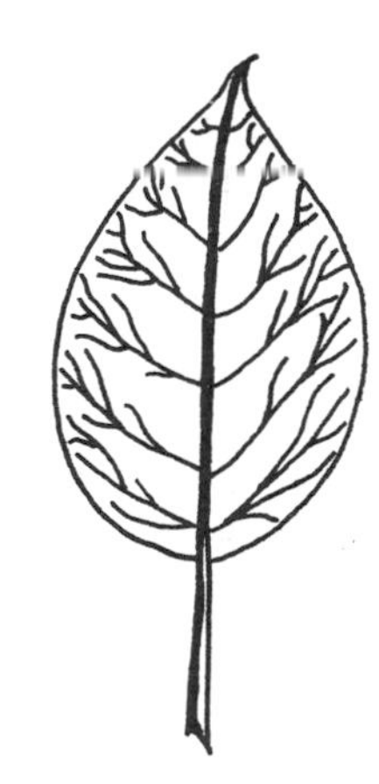

Fig. 5-7. Cross section of a leaf.

companied by internal secretion of resin. The resin-soaked area became yellow to pinkish as the cells died. Annual SO_2 concentrations as low as 0.03 ppm are thought to cause chronic pollution symptoms in plants. Some plants, like alfalfa and barley, are very sensitive, whereas other plants, such as chrysanthemums, are not particularly sensitive.

Control of sulfur dioxide. Since there is little that can be done about SO_2 pollution after it enters the air, some way is needed to prevent it from entering the air. The following approaches are available:

1. Cut down on use of the fuel.
2. Use sulfur-free or low-sulfur fuel.
3. Trap SO_2 before it enters the air.
4. Remove sulfur from the fuel (coal, oil, and gas).

The first approach is not possible, since we have insufficient nuclear, hydroelectric, solar, etc., sources of electricity. The second alternative is being used, but again the supplies of low-sulfur fuel are limited. In New York several years ago the fuel oil being burned averaged about 3% sulfur. Now only oil containing less than 1% sulfur may be used. As a result, the average hourly concentration of SO_2 had dropped from 0.21 ppm in 1964 to 0.11 ppm in 1969–a 50% reduction. The third alternative of trapping the SO_2 is being used by large industries and power companies today. Various types of "wet scrubbers" run the stack gases containing SO_2 through a solution such as sodium carbonate, where the SO_2 is absorbed. These devices are effective in industries that produce high concentrations of SO_2, such as coke production, but they are less effective in other industries where the concentration of SO_2 in the smokestack is relatively low. Electrostatic precipitators are effective in removing fly ash but do not remove the SO_2. Consequently, the absence of smoke from a smokestack does not mean that the effluent is clean. Unfortunately, all of these devices are expensive and therefore not practical for individual homes. The only solution available for home furnaces is to remove the sulfur from the fuel at the refinery, and this technique is now coming into practice. Recovered sulfur already amounts to over 50% of the free world's sulfur output.

NITROGEN OXIDES

When nitrogen in the air is subjected to very high temperatures such as those found in internal combustion engines, the nitrogen may react with oxygen to produce nitric oxide (NO). If the hot gases are allowed to cool slowly, the nitric oxide will decompose to again form nitrogen and oxygen. However, most engines are designed to quickly transfer the heat to some other function, such as expanding air to move a piston, allowing the gas to cool quickly. Under these conditions the nitric oxide remains intact; then the nitric oxide enters the air and begins to oxidize to form nitrogen dioxide (NO_2). Nitrogen dioxide is strongly absorbent in the blue-green portion of the visible spectrum, and this overbalances the yellow-red end of the spectrum and causes the sky to have a brownish color. Several other oxides of nitrogen are usually present in urban atmospheres, but they are of low toxicity. Moreover, they undergo reactions in the air in which the principal product is NO_2.

Nitrogen oxides in the environment. The most important problem associated with the nitrogen oxides is not the toxicity of the gases themselves, but the secondary pollutants that are produced when these oxides react with other chemicals in the air. The result is photochemical smog, which is discussed later in this chapter.

Up to 70% of various oxides of nitrogen are from automobile exhaust. High speed and rapid acceleration are responsible for most of these oxides. Consequently, the total amount produced in a given region is largely determined by the driving practices of the region. Exhaust emissions in typical urban traffic usually average about 1500 ppm total nitrogen oxides.

The odor threshold for pure NO_2 is 1 to 3 ppm. Eye and nasal irritation becomes apparent at concentrations of about 13 ppm, and accidental exposure to concentration of 150 to 200 ppm can be fatal. For unknown reasons intermittent exposure, such as for a worker in a factory, is much less harmful than continuous exposure to a similar concentration, such as breathing the city air, for the same total number of hours. Rats given continuous exposure to 5 ppm of NO_2 had 18% mortality, whereas those exposed intermittently to 5 ppm and 25 ppm for an equivalent number of hours had no mortality. Standards set by the Environmental Protection Agency require that air contain no more than 0.05 ppm of NO_2 on an annual basis. (Remember, man breathes this gas continuously!) Table 5-1 gives the monthly means for NO_2 in four United States cities. The global background level of NO_2 is 0.001 to 0.002 ppm.

In fog droplets the NO_2 can be converted to nitric acid as follows:

$$4NO_2 + 2H_2O + O_2 \longrightarrow \underset{\textbf{Nitric acid}}{4HNO_3}$$

This may take hours or days. The nitrogen oxides may switch back and forth to various compounds, but eventually they end up as nitrates, which are removed from the atmosphere by rain.

Mechanism of action. Unlike many other toxic materials, there is no observable response to NO_2 until a critical concentration is reached. There are

TABLE 5-1. Monthly mean levels (1967) of nitrogen dioxide (in ppm); federal regulations set 0.05 ppm as maximum allowable*

CITY	JAN.	FEB.	MARCH	APRIL	MAY	JUNE	JULY	AUG.	SEPT.	OCT.	NOV.	DEC.
Chicago	0.041	0.045	0.060	0.062	0.059	0.065	0.050	0.053	0.040	0.032	0.036	0.034
Cincinnati	0.029	0.029	0.036	0.028	0.026	0.026	0.027		0.034	0.026	0.026	0.026
Denver	0.038	0.032	0.042	0.024					0.041	0.038	0.045	0.048
Philadelphia	0.051	0.047	0.051	0.053	0.047	0.053	0.047	0.041	0.033	0.032	0.027	0.031

*Data from U. S. Environmental Protection Agency: Air quality data for 1969, Research Triangle Park, N. C., 1971.

usually no mild irritations of mucous membranes at presently encountered air pollution levels. At higher concentrations the gas can produce effects ranging from eye irritation, pulmonary congestion, and edema, to death.

Research on the biochemical effects of NO_2 has been largely concerned with enzymatic alteration. Exposure to NO_2 creates conditions that stimulate anaerobic oxidation (glycolysis) and inhibit aerobic oxidation (citric acid cycle and electron transport system). Nitrogen dioxide may also alter lung proteins. Inhalation of NO_2 can cause a change in the configuration of the structural proteins collagen and elastin. This has been demonstrated in rabbits after inhaling 1 ppm of NO_2 for 1 hour. This is a reversible change that may have some connection to emphysema. Along with the change in the structural protein molecules, there is also a disruption of cells in the lung tissue with a concomitant release of histamine. These changes are generally associated with inflammation. Reactions with lipids in the cell membrane may account for some effects of NO_2 inhalation.

Neither NO nor NO_2 is a serious problem for vegetation, since levels sufficient to cause injury are far above any known ambient levels. However, there is a significant decrease in growth when plants are exposed to 0.5 ppm of NO_2 for long periods of time.

Perspective. The pollution control devices now on automobiles are aimed at control of hydrocarbons. Unexpectedly, they have worsened the nitrogen oxide problem because the conditions that help to control the hydrocarbons tend to increase production of nitrogen oxides (see following section). The primary factors determining the rate of emission of nitrogen oxides from engines are the peak combustion temperature and the availability of oxygen at these high temperatures. Reducing either will reduce the nitrogen oxide emissions. One method of controlling nitrogen oxide emission is to reintroduce a portion of the exhaust gas into the cylinder. This tends to reduce the peak combustion temperature, since the inert exhaust gas serves as a heat sink, and the overall concentration of oxygen in the system is reduced. With this method it is possible to reduce emissions by as much as 90%, but it requires that up to 30% of the air taken into the combustion chamber must be recycled exhaust gas. Unfortunately, the inert exhaust gas decreases the power produced by the engine.

Our present understanding of the problem indicates that it may be possible to reduce nitrogen emissions from internal combustion engines to 100 ppm from the present average level of about 1500 ppm.

OXIDANTS AND PHOTOCHEMICAL SMOG

Photochemical smog ("Los Angeles smog," Fig. 5-8) is largely the result of incomplete combustion of gasoline in the automobile internal combustion engine. Two of the major components of auto emissions are (1) unburned hydrocarbons, which result from incomplete combustion of gasoline, and (2) nitrogen oxides, which are formed when nitrogen and oxygen react at the high temperatures in the combustion chamber. Since its beginning, control of photochemical smog has been oriented toward reduction of hydrocarbon emissions, and very little has been done to control the other major reactants—the nitrogen oxides—which play a special role in the photochemistry of the air.

Before a photochemical reaction can occur in the atmosphere light must be absorbed by the reacting atoms or molecules. Unlike most air pollutants, one of the oxides of nitrogen (NO_2) is brown colored, and this permits it to capture the energy of the sunlight (especially the ultraviolet radiation) and start a series of reactions that produce new (secondary) pollutants known as *oxidants*. Ozone (O_3) is the principal oxidant in photochemical smog, but in recent years another group of oxidants, the peroxyacyl nitrates (PAN compounds), has been found to be important as well. Not only are these oxidants highly toxic but they have a great reactive potential.

Fig. 5-8. Los Angeles on a typical workday afternoon. The photochemical smog increases in density throughout the day. (Courtesy Los Angeles County Air Pollution Control District, Los Angeles, Calif.)

They react with unburned hydrocarbons in the air to produce new and different toxic and irritating air pollutants.

Photochemical reactions. When nitric oxide (NO) is released from an engine, it reacts very slowly with oxygen to form nitrogen dioxide (NO_2).

$$2NO + O_2 \longrightarrow 2NO_2$$

However, this conversion is much more rapid when ozone is present in the atmosphere. Without ozone the reaction may take several days, but it takes only a few seconds when ozone is present. With ozone the reaction is as follows:

$$NO + O_3 \longrightarrow NO_2 + O_2$$

In the presence of sunlight NO_2 is split to form NO and atomic oxygen (O), but this reaction is reversible and some of the atomic oxygen reacts with NO to again form NO_2 as follows:

$$NO_2 \rightleftharpoons NO + O$$

This means that once NO_2 is formed, it can continue to regenerate itself in the atmosphere, but as long as sunlight is present, atomic oxygen will continue to be produced. This dissociation of NO_2 into NO is probably the most important primary photochemical process. Atomic oxygen is also reactive and unites with molecular oxygen (O_2) to form ozone (O_3), the principal oxidant in photochemical smog.

$$O + O_2 \longrightarrow O_3$$

Ozone is normally present in the atmosphere at levels of 0.01 to 0.03 ppm, an amount not detectable by the human senses. It becomes irritating at about 0.2 ppm and a health menace at 1 ppm. Levels of 0.99 ppm have been measured in Los Angeles. The highly reactive ozone acts as an oxidizing agent and triggers a vast number of reactions involving other atmospheric contaminants. Included among the more important of these contaminants are the unburned hydrocarbons from automobile exhaust.

Most organic substances oxidize, or break down, with time. Foods deteriorate, rubber disintegrates, petroleum products become gummy, and colors fade. All of these changes are due to chemical reactions involving oxidation, which is the addition of oxygen or removal of hydrogen from a molecule. The presence of ozone greatly accelerates the rate at which these reactions occur. Following is a sample reaction involving ozone and ethylene (a hydrocarbon found in gasoline). When unburned ethylene from exhaust comes in contact with the ozone formed in the atmosphere, the molecules react producing various *parts* of complete molecules called free radicals and formaldehyde.

$$\underset{\text{Ozone}}{O_3} + \underset{\text{Ethylene}}{H_2C{=}CH_2} \rightarrow \underset{\text{Formaldehyde}}{H_2C{=}O} + \underset{\text{Free radical}}{\dot{H}O} + \underset{\text{Free radical}}{H\dot{C}O}$$

These free radicals (symbolized by a dot over that piece of the original molecule) can be thought of as chemical units "just waiting to undergo a chemical reaction." Almost any organic compound that absorbs solar radiation can produce free radicals. For example, when a molecule such as acetone (fingernail polish remover) evaporates and is hit by sunlight, the following reaction can occur:

$$\underset{\text{Acetone}}{(CH_3)_2C{=}O} \xrightarrow{\text{Sunlight}} \underset{\text{Free radical}}{\dot{C}H_3} + \underset{\text{Free radical}}{CH_3\dot{C}{=}O}$$

These free radicals readily combine with oxygen in the atmosphere to produce peroxyl radicals as follows:

$$\underset{\text{Free radical}}{\dot{C}H_3} + O_2 \rightarrow \underset{\text{Peroxyl radical}}{CH_3O\dot{O}}$$

To continue this particular example the peroxyl radical may then react with nitrogen oxides or other air pollutants to form a great variety of compounds, including the peroxyacyl nitrate (PAN) oxidants. The following outline, starting with just one organic compound, shows what can happen when the inorganic and simple organic components of automobile exhaust are subjected to solar radiation.

$$CH_3O\dot{O} + \left\{ \begin{array}{l} NO \\ NO_2 \\ C_nH_{2n-1}R \\ SO_2 \\ O_3 \end{array} \right\} \rightarrow \left\{ \begin{array}{l} \text{Alkyl nitrates} \\ \text{Peroxyacyl nitrates (PAN)} \\ \text{Alcohols} \\ \text{Ethers} \\ \text{Acids} \\ \text{Peroxyacids} \end{array} \right.$$

These reaction products are also subject to chemical and photochemical attack. For example, photochemical breakdown of a peroxyacyl nitrate can result in the formation of acylate radicals and NO_2. Considerable research is being done to try to determine the complex reactions occurring in photochemical smog.

Effects of oxidants on organisms. Of all the vast array of complex and toxic compounds that can result from photochemical reactions, two are particularly hazardous (ozone and peroxyacyl nitrates, Fig. 5-9). These two compounds are abundant in the Los Angeles type of smog as opposed to a London type of smog, where carbon monoxide, sulfur dioxide, and nitrogen dioxide are the major pollutants.

Ozone, a major component of photochemical smog, is also present (0.02 ppm) in "pure country air." Many people associate ozone with pure air. In fact, many home air cleaners use ozone as a purifier. These devices may add to the ozone level already considered to be too high. The "safe level" for total photochemical oxidants, of which ozone comprises about 90%, is set at 0.08 ppm. Smog irritation is detectable at 0.15 ppm and clearly noticeable at 0.25 ppm. Levels of 1 ppm have been measured in Los Angeles, but this problem is not restricted to that city. Oxidant concentrations of 0.32 ppm have been measured in Cincinnati and other cities.

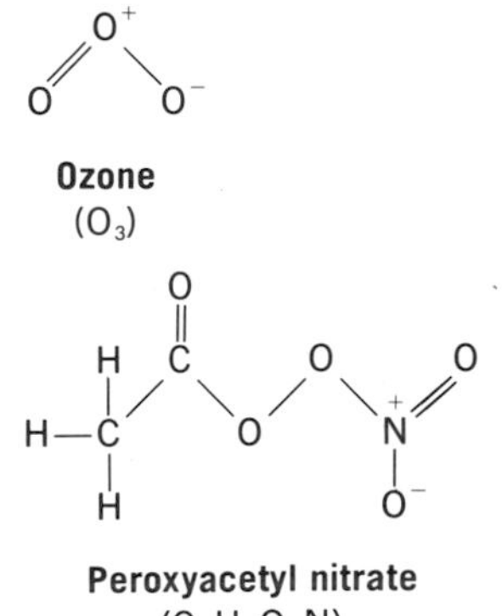

Ozone is a highly reactive gas and is unlikely to penetrate much beyond the respiratory tract. The main effects are on the respiratory tract, therefore, but it also causes drowsiness and headaches, eye irritation, and changes in visual fields and night vision. Animals also show premature aging when exposed to ozone. It has been suggested that many ailments, including lowered resistance to pulmonary infections, may be brought about by the levels of atmospheric ozone occasionally found in many cities.

In plants the initial effect is in the palisade layer and involves scattered groups of cells. The leaves develop brown flecks or stippling on the upper leaf surface, particularly in the older leaves on the plant (Fig. 5-10). Younger leaves seem to be more resistant. The cells plasmolyze (lose water) and disintegrate. After continued exposure the dead areas on the upper surface get larger and coalesce. Eventually the spongy cell layer is damaged. Although the ozone enters through the stomata on the bottom surface of the leaf, the palisade layer is attacked first. The reason why this happens is not known. Some plants, such as the Virginia pine, European larch, and eastern hemlock are susceptible to ozone, whereas others such as the balsam fir, red pine, and white spruce are resistant. In onions, ozone resistance is regulated by dominant genes, possibly a single pair of genes. The dominant trait causes the stomata of the resistant variety to close on exposure to ozone, whereas the stomata of the sensitive variety remain open.

Ozone
(O_3)

Peroxyacetyl nitrate
($C_2H_3O_5N$)

Fig. 5-9. Principal oxidants in photochemical smog. Ozone is the most abundant oxidant. Peroxyacetyl nitrate is one of the many PAN-type oxidants (peroxyacyl nitrate compounds).

Fig. 5-10. Oxidant injury to grape leaves. Damage is indicated by mottling and yellowing, illustrated here by dark flecks and light-colored leaves. Areas of extreme damage are circled. (Courtesy Environmental Protection Agency, Research Triangle Park, N. C.)

Only in the last few years have the PAN-type compounds been identified and recognized as part of the smog problem. These compounds are not as toxic as ozone but seem to be about as poisonous as nitrogen dioxide. However, other nonlethal effects are important. A 12-minute exposure to 0.5 ppm of PAN causes eye irritation. At 2 ppm of PAN a 5-minute exposure causes significant irritation. Tests with PAN have shown that concentrations as low as 4 ppm reduce spontaneous activity in mice by 50%.

PAN compounds cause a characteristic type of plant injury. Actually, other PAN-type compounds such as peroxypropionyl nitrate (PPN) and peroxybutyryl nitrate (PBN) are more toxic than PAN, but PAN is the most abundant of these compounds. PAN affects the lower surfaces of young leaves and the younger tissues of older leaves. It causes the spongy parenchyma cells near the stomata to collapse. This produces a characteristic silvering or bronzing of the leaf undersurface, which ruins their sale as leaf vegetables. Usually the upper leaf surface is damaged only after the lower surface has extensive damage. In other cases damage may initially extend through the entire thickness of the leaf.

Mechanism of action. Extensive research has been done on the effects of air pollution on the respiratory system, and one of the important findings is that ozone affects macrophage cells. These cells, found in many parts of the body, have the ability to ingest almost any kind of foreign particles, including bacteria. Any change in the function of pulmonary macrophage activity by a pollutant could have serious effects on the ultimate health of the animal. In addition to handling inhaled particles they also are involved in the initiation of antibody synthesis. Low levels of ozone have been shown to decrease the activity of various enzyme systems (acid phosphatase, beta-glucuronidase, lysozyme) in rabbit lung macrophages. In a related study increased numbers of bacteria were consistently cultured from the lungs of mice exposed to ozone as compared to unexposed mice, and the magnitude of the increase in bacterial numbers correlated with increases to ozone exposure up to 2.58 ppm. This was apparently a consequence of the toxic effect of ozone on the alveolar macrophage cells. Bronchial mucus normally contains large quantities of the antibacterial enzyme lysozyme, but mice and rabbits exposed to ozone show reduced amounts of this enzyme. Twelve hours after exposure the enzyme activity returns to normal levels. The present theory is that loss of the lysozyme activity by alveolar cells is due to oxidation of the lysozyme by ozone. Ozone and PAN can oxidize the reduced form of nicotinamide adenine dinucleotide (NAD). The mechanism of action of these oxidants at the cellular and subcellular levels may be an interference with the electron transfer systems in biological oxidations (Chapter 2).

Benzo[*a*]pyrene, a carcinogen associated with coal and cigarette smoke or smoke from charcoal-broiled steaks, is rendered less harmful by an enzyme called benzo[*a*]pyrene hydroxylase. However, exposure to increasing ozone concentrations can decrease the activity of benzo[*a*]pyrene hydroxylase by as much as 70%. Therefore ozone may act as a cocarcinogen with inhaled benzo[*a*]pyrene by delaying the enzymatic transformation of the primary carcinogen benzo[*a*]pyrene.

Perspective on air pollution. What people are learning about air pollution and its effects does not fit into any simple system from which any firm conclusions can be drawn. Even attempting to set the limits of "safe" concentrations of air pollutants is difficult. One problem associated with setting these standards is the long period between initial exposure and the resulting ill effects. The initial symptoms may appear in a few days, but some effects may not be seen for 30 or 40 years. Another unanswered question is the relationship between the concentration of the pollutant and the length of exposure. Certainly no one would argue that exposure to 1 ppm of nitrogen dioxide for 365 days would have the same effects as exposure to 365 ppm of nitrogen dioxide for 1 day. What is the effect of a lifetime of exposure to "safe" levels of air pollutants? Does air pollution have the same effects during the lifetime of a laboratory animal (3 to 4 years) as it does on humans with a life-span of 75 years?

Still unanswered is the question of synergistic effects of pollutants. That is, are the combined effects of several pollutants added or are they multiplied? Compound A by itself may be considered safe, and compound B by itself may be considered safe, but when compounds A and B are present together, they may interact and produce a much different effect than either produced separately. In most laboratory studies so far only the effects of one or two materials have been investigated at a time (e.g., SO_2 and NO_2 or CO and SO_2). Compared to individual pollutants, studies of synergistic effects of several pollutants are more difficult to conduct and much more costly, and they are even harder to interpret.

In short, it will be a long time before there are answers to some of the most important questions about the effects of various air pollutants on living things.

CYANIDE

Almost everyone is aware that cyanide is a poison. Most people have heard of it as the "active ingredient" in gas chamber executions. What most people do not realize is that cyanide is a common water pollutant. Small quantities of cyanide do occur naturally, such as that produced by millipedes as a defense mechanism against their predators. However, all of the cyanide in industrial use today is commercially synthesized. Cyanide consists of carbon and nitrogen atoms joined by a triple bond ($C \equiv N$) to form the cyanide radical. Compounds containing the cyanide radical (e.g., sodium cyanide, hydrogen cyanide) form a class of versatile reagents with many chemical and industrial applications. These chemicals enter industrial waste streams from a variety of chemical processing industries such as extraction of gold and silver ores, synthetics manufacturing, coal-coking furnaces, and electroplating of gold, silver, zinc, cadmium, and other metals.

Mechanism of action. Classed as a respiratory inhibitor, cyanide severely reduces cell respiration by binding irreversibly to the iron group of cytochrome a+a_3, an enzyme complex necessary for the final energy step of the electron transfer system in mitochondria (Chapter 2). Fig. 5-11 shows the point of cyanide blockage. Cyanide binding of cytochrome a+a_3 reduces the oxygen intake of cells immediately, and this effect is particularly damaging to nerve cells. Symptoms of cyanide poisoning are rather extreme: first the nervous system is excited, then paralyzed; the pupils are constricted, then dilated; simultaneously respiration is increased, then abruptly halted. Since cyanide binds irreversibly, extremely low concentrations can prove fatal to any living organism; for example, concentrations greater than 0.1 ppm of cyanide in water can kill fish.

Control of cyanide pollution. Because cyanide concentrations as low as 0.3 ppm are toxic to bacteria in activated sludge, municipal sewage treatment facilities often place strict regulations on the cyanide content of effluents entering sewer systems. Industries using cyanide for processing are thus required to

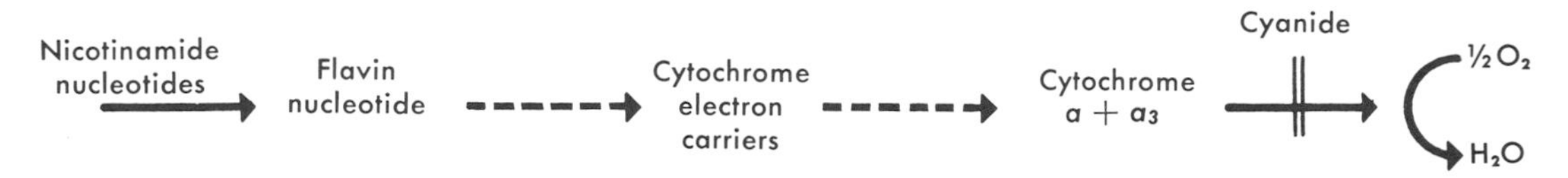

Fig. 5-11. Electron transfer system in mitochondria, showing where cyanide inhibits the reaction. For further information on the electron transfer system see Chapter 2.

use available on-site cyanide destruction processes. Fortunately, there exist some effective techniques for removing cyanide compounds from industrial effluents.

Detoxification of cyanide involves rupturing the triple bond between the carbon and nitrogen atoms (C≡N). The most frequently used method is the alkaline chlorination process in which sodium hypochlorite (household bleach) is added directly to waste water, where it rapidly oxidizes the cyanides to less toxic cyanates. Cyanates, although less toxic than cyanides, are unacceptable wastes to some authorities and must be further converted to harmless carbon dioxide and ammonia by treatment with strong acid. New catalytic chemical processes have been developed and are being put into operation by some industries to meet stringent water quality regulations. Although these new processes employ slightly different chemical pathways, they ultimately detoxify cyanide by breaking it down into carbon dioxide and ammonia.

SUGGESTED READING

Aldrich, S. R.: Some effects of crop-production technology on environmental quality, Bioscience **22**:90-95, 1972.

Eisenbud, M., and Ehrlich, L. R.: Carbon monoxide concentration trends in urban atmospheres, Science **176**:193-194, 1972.

Heller, A., and Ferrand, E.: Low sulfur fuels for New York City. In Hart, J., and Socolow, R., editors: Patient earth, New York, 1971, Holt, Rinehart & Winston, Inc.

Jaffee, L. S.: The biological effects of photochemical air pollutants on man and animals, American Journal of Public Health **57**:1269-1277, 1967.

Keeney, D. R.: The fate of nitrogen in aquatic ecosystems, Madison, Wis., Jan., 1972, University of Wisconsin Water Resources Center.

Kellogg, W. W., and others: The sulfur cycle, Science **175**:587-596, 1972.

Maugh, T. H.: Carbon monoxide: Natural sources dwarf man's output, Science **177**:338-339, 1972.

Plass, G. N.: Carbon dioxide and climate, Scientific American **20**(1):41-47, 1959.

Rasool, S. I., and Schneider, S. H.: Atmospheric carbon dioxide and aerosols: Effects of large increases on global climate, Science **173**:138-141, 1971.

Rukeyser, W. C.: Fact and foam in the row over phosphates, Fortune **85**(1):71-73, 166-170, 1972.

Stern, A. C.: Air pollution, vols. I to III, New York, 1968, Academic Press, Inc.

Villiers, A. J.: The effects of air pollution on health, Occupational Health Review **20**(3-4):25-44, 1968-1969.

Wolff, I. A., and Wasserman, A. E.: Nitrates, nitrites, and nitrosamines, Science **177**:15-19, 7 July 1972.

6 INDUSTRIAL AND MUNICIPAL ORGANIC COMPOUNDS

This chapter will be concerned with the majority of natural and man-made organic compounds that have proved troublesome in the environment. The chemist defines *organic* compounds as those that contain carbon in their structure. In the previous chapter several simple carbon compounds were considered. Those to be examined in this chapter are all of a more complicated structure, containing several to many carbon atoms per molecule. The pesticides, because of their particular importance and widespread use, will be treated separately in Chapter 7. With the exception of oil, all of the compounds to be discussed are man made. In the case of the combustion hydrocarbons some of the compounds also result from natural forest fires, but quantitatively those that are man generated are more important. The only thread of chemical continuity that runs through this chapter is the presence of carbon in the structure of all of the molecules taken up. Similarly, the effects of the various organic molecules on living organisms are almost as varied as the materials themselves. In short, each section of this chapter can be considered as largely independent of the others. In most cases the mechanisms through which the organic pollutants produce their effects on organisms remain totally obscure. That these mechanisms are not as well understood as those of the elemental and inorganic pollutants is not surprising in view of the complex nature of the reactions of these larger molecules. Toxicologists are only beginning to study these reactions, but the day will come when they will be able to explain how hexachlorophene, for example, inflicts its brain damage on mammals. Until then, descriptions of the symptoms of poisoning by the various compounds must suffice, and attempts must be made to find ways to control their release into the environment.

COMBUSTION HYDROCARBONS

Description, sources, and occurrence. In any listing of air pollutants there is a category labeled hydrocarbons. In 1970 this category accounted for about 13% of the total United States air pollution by weight. The term *hydrocarbons* encompasses a vast array of organic compounds, most of which remain unidentified. All of the common categories of organic molecules (e.g., alcohols, aldehydes, ketones) are represented, together with an assortment of more complex compounds like benzo[*a*]pyrene. Most of these chemicals come from various combustion processes and therefore are termed *combustion hydrocarbons.* The remainder are by-products of oil refining and other industrial processes (p. 134). The combustion hydrocarbons enter the atmosphere as a result of incomplete burning of organic materials (Fig. 6-1). This kind of combustion occurs readily during open-air burning, in internal combustion engines (particularly when they are worn or improperly adjusted), and in simple forms of space-heating devices. It does not happen to any great extent in most industrial combustion devices (e.g., industrial boilers, power

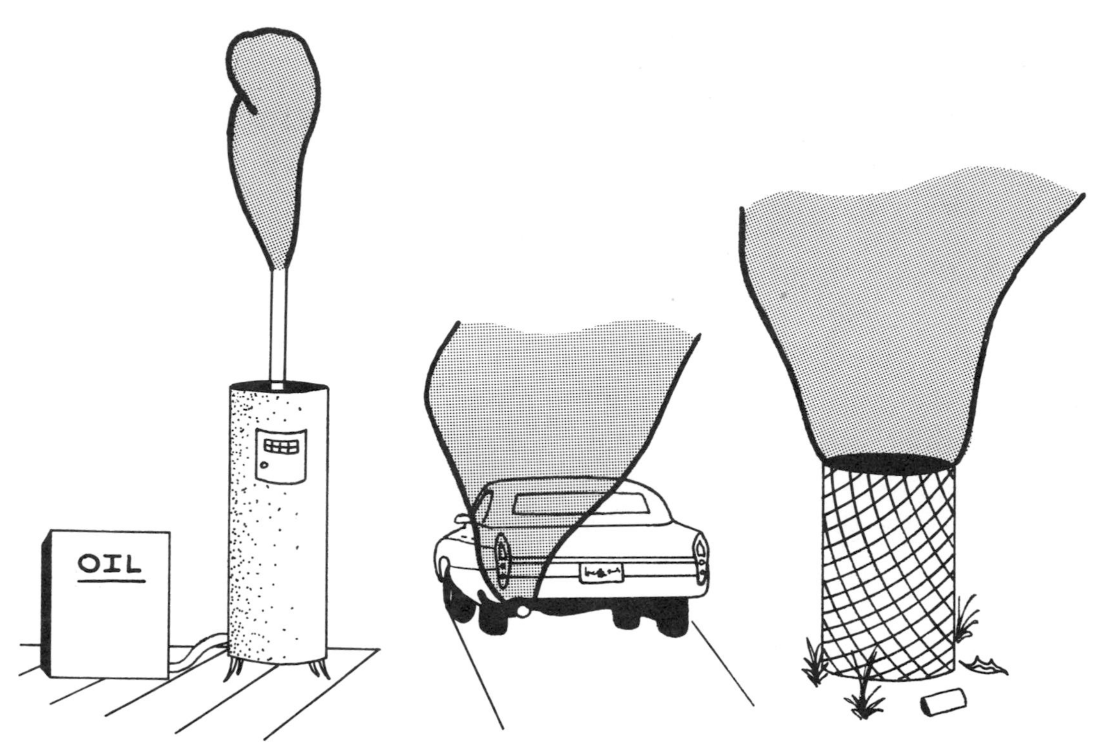

Fig. 6-1. Primary sources of combustion hydrocarbons: oil-fired space heaters, the incomplete combustion of automotive fuels, and open burning of trash.

TABLE 6-1. Estimated emissions of hydrocarbons in United States in 1970*

SOURCE	MILLIONS OF TONS
Transportation	19.5
Fuel combustion in stationary sources	0.6
Industrial processes	5.5
Solid waste disposal	2.0
Miscellaneous	7.1
Total	34.7

*From U. S. Council on Environmental Quality, Third Annual Report, Washington, D. C., 1972, U. S. Government Printing Office.

plants, or municipal incinerators) because such incomplete burning of fuel constitutes an economic loss that cannot be tolerated. Table 6-1 summarizes the estimated hydrocarbon emissions from the most important sources.

Effects on organisms. The role of combustion hydrocarbons in smog formation and the subsequent effects of photochemical oxidants, etc., on organisms have been described earlier (Chapter 5). In this section the direct effects of the hydrocarbons on organisms will be discussed. These effects can be divided into acute and chronic categories. The acute effects, which primarily take the form of irritation of the eyes, nose, and throat, can be caused by relatively low levels of many of the combustion hydrocarbons.

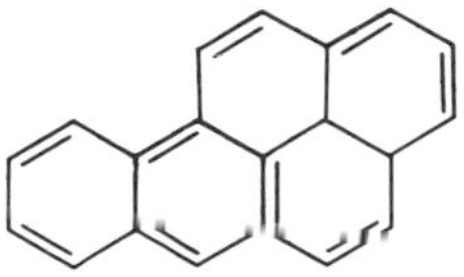

Benzo[a]pyrene

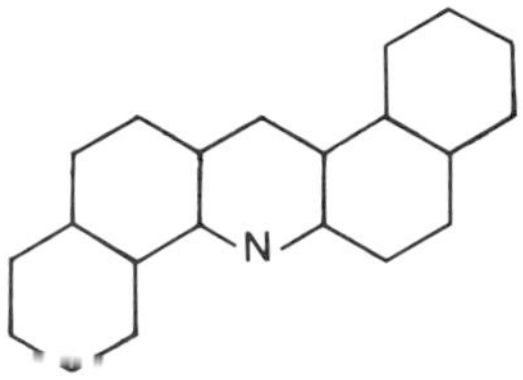

Dibenz(a,h)acridine

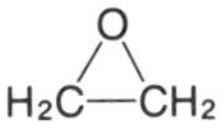

$H_2C—CH_2$ (with O bridging)

Ethylene oxide
(an epoxide)

Formaldehyde causes eye and throat irritation and inhibition of ciliary action in the respiratory passages at levels of 0.2 to 1 ppm. Some of its chemical relatives like acrolein and crotonaldehyde are much more irritating to mucous membranes than formaldehyde itself, showing similar effects at even lower concentrations. Eye irritation, largely from this group of compounds, is one of the most frequent complaints in areas of severe air pollution. The inhibition of ciliary activity by exposure to formaldehyde vapor is important in consideration of the chronic effects of the combustion hydrocarbons.

The chronic effect of greatest concern from this group of chemicals is cancer induction. Three groups of combustion hydrocarbons have been implicated in carcinogenesis. These are the polynuclear aromatic hydrocarbons like benzo[*a*]pyrene (BaP or 3,4-benzpyrene), the polynuclear heterocyclic compounds like dibenz(a,h)acridine, and the alkylating agents, which include epoxides, peroxides, and lactones (see above). At least some members of all of these groups have been shown to cause cancer in laboratory animals. However, there is no irrefutable evidence of any air pollutant having caused cancer in a human being. Circumstantial evidence linking these kinds of compounds with human cancer is very strong, but the evidence needed to prove a cause-and-effect relationship beyond doubt is difficult to get. Most of the research on the carcinogenicity of combustion hydrocarbons has centered around the polynuclear aromatic compound BaP. This compound has received a great deal of attention because some years ago it was shown to be an active, tumor-inducing compound and also because it is one member of this group that is quite easy to detect and measure in ambient air. Consequently, data on the concentration of BaP in the effluent of many sources and in the air of major cities are readily available (Tables 6-2 and 6-3). The most important producer of BaP is the inefficient combustion of coal, as is frequently found in homes and small industrial coal-fired furnaces. Larger users of coal burn it much more completely so the quantity of BaP emitted is much less than that from the smaller sources, even though the amount of fuel consumed by the large industries is much greater. Motor vehicles also contribute to the total BaP produced, as do refuse burning and various industrial processes.

The effects of BaP on experimental animals are now well documented. The methods of exposing the experimental animals to this carcinogen differ from those found in natural conditions, but this is frequently the case in animal toxicity studies and should not invalidate the data obtained. Lung cancer has been induced in a variety of animals by administration of BaP adsorbed on iron particles, BaP in polysorbate 80, and "artificial smog" (ozone-treated gasoline vapor). Experimentally the chemical was usually injected into the trachea rather than through natural inhalation. In connection with these studies it was found that the addition of one, two, or three methyl groups ($-CH_3$) to the BaP molecule enhanced its carcinogenicity. The enhancing effects of iron oxide particles used as "carriers" for the BaP mole-

TABLE 6-2. Sources of benzo[*a*] pyrene (BaP) emissions; estimated annual BaP emissions for the United States*

SOURCE	TONS	PERCENT OF TOTAL
Heat generation		
Residential	410	85
Commercial and industrial	9	1.9
Electric power generation	3	0.6
Refuse burning		
Incineration	5	1.0
Open burning	15	3.1
Industrial processes	19	3.9
Motor vehicles		
Autos	9	1.9
Trucks and buses	12	2.5
Total	482	99.9

*Data from Hangebrauck, R. P., vonLehmden, D. J., and Meeker, J. E.: Sources of polynuclear hydrocarbons in the atmosphere, pub. no. 999-AP-33, 1967, U. S. Department of Health, Education and Welfare.

TABLE 6-3. Average levels of benzo[*a*] pyrene (BaP) in the air of selected United States locations, 1967*

LOCATION	CONCENTRATION (ng/m^3)
Los Angeles, Calif.	1.31
Chicago, Ill.	2.99
Indianapolis, Ind.	5.67
Kansas City, Mo.	7.07
Altoona, Pa.	29.53
Washington, D. C.	1.88
Grand Canyon National Park, Ariz.	0.18
Yellowstone Park, Wyo.	0.07

*From U. S. Environmental Protection Agency: Air quality data for 1967, Washington, D. C., 1971, U. S. Government Printing Office.

cules are believed to be caused by the iron oxide particles penetrating the alveolar walls and carrying the BaP to the sensitive lung tissues. The iron particles appear to be engulfed by the macrophages and hence remain in contact with the lung tissue for an extended period of time. This keeps the BaP in the lung, where it can accumulate to high levels, increasing the likelihood of tumor formation. A similar mechanism has been hypothesized for the lung cancer–inducing mechanism involved in asbestos-induced cancers. Some types of asbestos contain oils with a high content of BaP, reaching 1 to 5 μg per 100 grams of asbestos.

SOLVENTS

Description, occurrence, and uses. Organic solvents are among the most widely used industrial materials. Examples of this group include alcohols, ketones, and low molecular weight aromatic compounds like benzene and toluene, as shown on opposite page. They are found in many manufactured products such as paints, varnishes, and printing inks, but probably more important from a pollution standpoint, they are used in a great variety of industrial operations. As an example, metal parts must be cleaned and degreased before they can be plated. This process requires a fat solvent (e.g., methyl ethyl ketone) in which the parts can be washed. The rate of solvent evaporation from

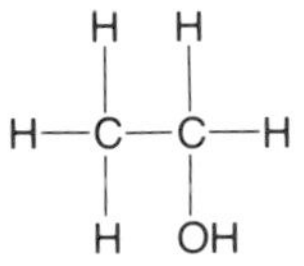

Ethyl alcohol 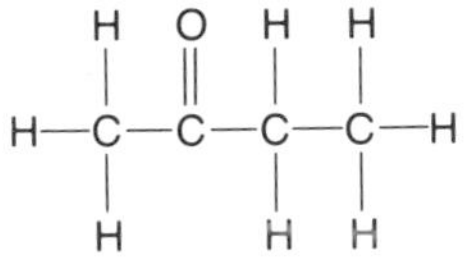   Methyl ethyl ketone 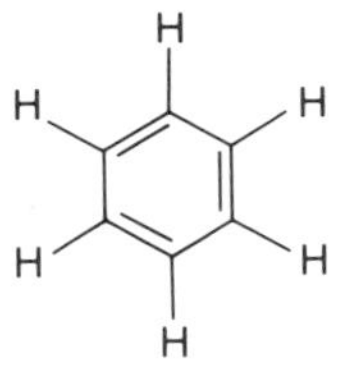   Benzene 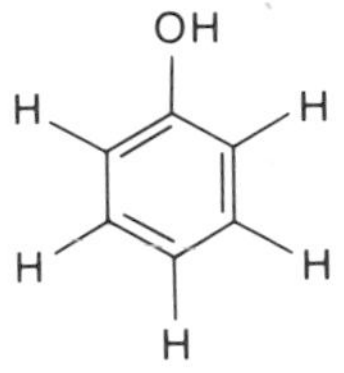   Toluene

such washing operations is high, and the resultant air pollution is of major concern. Another important problem is that solvent solutions eventually become loaded with contaminants and consequently become ineffective and have to be disposed of. The simplest way for the plant operator to get rid of the spent solvent is to flush it down the drain, but it immediately becomes a water pollutant when he does so. The solvents are also important causative agents in industrial odor complaints–an area of environmental problems that is only beginning to receive attention.

Effects on organisms. As just mentioned, solvents are of concern both as air and as water pollutants. In the atmosphere they contribute to photochemical smog formation in the same way as do hydrocarbons from other sources. In the water they are toxic to all forms of aquatic life, although different species have widely different thresholds of resistance. Only rarely do humans encounter these compounds in sufficient concentrations to cause direct injury, but it does happened occasionally. One reported incident of solvent poisoning involved a group of Japanese women working in a small manufacturing plant that used toluene as a solvent. The women lived in an apartment on the third floor of the factory building and encountered toluene vapor at a concentration of 30 ppm in their quarters. In the work area of the plant toluene was present in the air at 150 to 550 ppm. Three of four women that had worked and lived in the building for 1 to 10 years developed aplastic anemia, a defect in which the bone marrow ceases to form normal blood cells.

Water pollution by solvents is a problem in all industrial areas. Sewage treatment plant operators report problems when these materials arrive in the normal municipal waste stream. The bacteria that are required for secondary sewage treatment are sensitive to a variety of industrial chemicals, including solvents. On occasions, when particularly massive amounts of such chemicals are allowed to flow into the municipal sewage system, they can cause a nearly complete kill of the desirable bacteria. When this happens, the treatment plant must allow the waste stream to pass through with only primary treatment until the toxic material can be flushed out and bacterial populations reestablished. Because of this continuing problem, most municipalities have established requirements for on-site treatment of industrial discharges to remove the solvents and other toxic wastes before the plant effluent is dumped into the municipal sewage system.

Perspective. The national ambient air quality standard for hydrocarbons set by the Environmental Protection Agency in 1971 is 160 $\mu g/m^3$ of air. This standard was set with the role of hydrocarbons in smog production in mind, not because of direct effects of hydrocarbons at this level. Achieving this standard will require control of solvent vapor emissions, at least from the larger sources. Technology for accomplishing this control is available and is being applied. Many industries are taking steps to eliminate all solvents from their liquid wastes by collecting them and then incinerating them in specially designed burners that release essentially nothing except carbon dioxide and water vapor. Eliminating the relatively minor amounts of solvent vapor that are responsible for the smells associated with solvent-using facilities will be much more difficult, and from a health standpoint it is a much lower priority item.

OIL

Description, occurrence, and uses. In the context of environmental issues the term *oil* refers to crude petroleum (as it comes from the ground) or to any of its refined by-products. Oil, in this sense, is not a single substance but one of the most complex and variable mixtures known to man. It contains many different aromatic and aliphatic* compounds with small but important amounts of a number of contaminants (e.g., sulfur, oxygen, nitrogen, various metals) included. The exact composition of any particular oil deposit, commonly called its "chemical fingerprint," can be determined by gas chromatography–mass spectroscopy. This technique provides the possibility of identifying the source of any particular oil spill. There are few uses for crude oil itself, but it is refined into a vast array of products from liquefied petroleum gas (LPG) to paraffins and tars. The refining process consists mainly of separating the various components of crude oil by distillation. This process is augmented by "catalytic cracking," a procedure in which a metallic catalyst is used in the breakdown of larger molecules into smaller ones. The cracking process increases the percentage of low molecular weight (hence more volatile) molecules that are used as fuels. Many of the separate fractions are subsequently recombined to yield products with particular sets of properties.

Oil occurs in the form of underground pools that are tapped by drilling, the oil being pumped to the surface. From the wells it is transported to refineries, either by pipelines or ships, where it is converted into useful products. Ideally the production-transportation-refining-use-disposal sequence constitutes a closed system with no escape to the environment. In practice this is not the case, and at every step along the line, but particularly during transport, significant quantities of petroleum products are leaked into the environment. Collectively the large and small leaks, regardless of the place in which they occur in the use cycle, disrupt the normal functioning of the world's ecosystems.

Current annual petroleum production stands at somewhat over 2 billion tons, and estimates are that this figure will more than double by 1980. More than 60% of the total oil produced is transported by ocean-going tankers (oil cargoes approximately equal the tonnage of all other types). Losses from these tankers constitute the most significant source of oil pollution. The wreck of the supertanker *Torrey Canyon* off the coast of England in 1967 spilled over 100,000 tons of oil into the sea. The story of this incident made headline news around the world because of the size of the tanker involved and the amount of oil spilled, but people seldom hear about the other tanker groundings that occur on an average of once a week somewhere in the world. The new generation of supertankers has ships that are twice as large as the *Torrey Canyon,* and tankers five times her size are on the drawing boards, raising the specter of 500,000-ton spills in the near future (Fig. 6-2). A spill of this amount could create a slick ¼ inch thick over more than *30 square miles* of sea surface, or coat 120 miles of beach with a layer of oil 1 inch thick and 100 yards wide. It has been estimated that from 5 to 10 million tons of oil are being added to the oceans every year. This figure includes both natural and man-made sources, but it has been calculated that natural leaks probably account for only a small fraction of the total. Man-caused spills are not randomly distributed over the surface of the seas but, rather, occur along a few well-defined shipping lanes

*Aromatic compounds are those that contain the six-carbon benzene ring

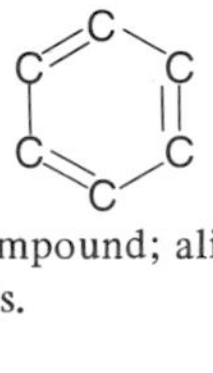

C C C C C C

or are related to this compound; aliphatic compounds are all other organic compounds.

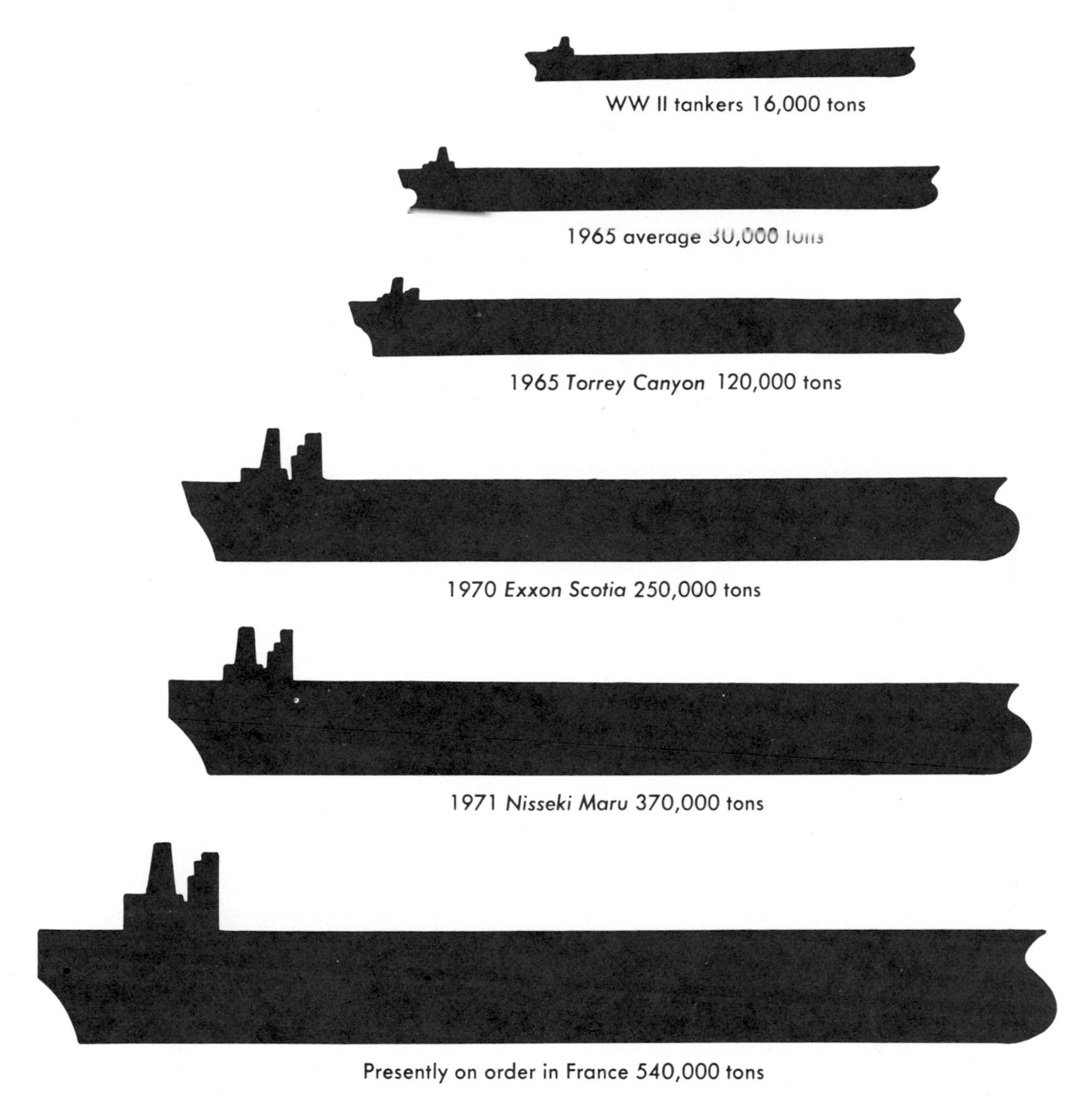

Fig. 6-2. Increase in capacity (dead weight tons) of oil tankers since World War II.

(e.g., 25% of all the oil shipped passes through the English Channel). Most spills occur near land in shallow inshore waters—the same waters that contribute most of the food currently being harvested from the sea. Here the biological effects of a spill will be much greater than if an equal amount of oil had been dumped into the open ocean. Another factor entering into the biological-effects equation is that approximately one half of the oil tonnage lost from ships consists of *refined* oil, which is generally considered to be more toxic than the crude oil from which it is derived.

Effects on organisms. Crude oil is a complex mixture of hydrocarbon compounds of varying toxicity. Four basic types of hydrocarbons are found in the typical crude oil: low boiling (low molecular weight) saturated hydrocarbons, high boiling (higher molecular weight) saturated hydrocarbons, low boiling aromatic compounds, and high boiling aromatic compounds. An additional group, the unsaturated hydrocarbons, or olefins, are not usually present in crude oil but make up a significant part of many refined petroleum products. For a long time the lower molecular weight fractions (both aromatic and aliphatic) have been considered of minor importance in oil spill accidents because of the general belief that they would evaporate into the atmosphere before causing serious harm to aquatic organisms. However, recent evidence indicates that this evaporation process is not nearly complete and that a significant proportion of these compounds remains in and on the water for days or weeks after a spill. The low boiling saturated compounds have been shown to produce anesthesia and cell death in a variety of marine organisms, with larval forms being particularly sensitive. The low boiling aromatic fraction of crude oil is probably its most dangerous component. It contains the familiar solvents benzene, toluene, and xylene, all of which are acute poisons to man as well as other animals. Much of the damage to marine life occurring after the *Torrey Canyon* spill could be traced to this class of compounds. Ironically most of these were intentionally added to the sea to dissolve the 2.5 million gallons of detergents that were used to clean up the beaches. In retrospect the detergents probably caused more damage to marine life than the oil itself.

The high boiling aromatic compounds include many complex four- and five-ring molecules that are suspected of being long-term poisons. Some of them, such as BaP, have been demonstrated to be carcinogenic in test animals. The higher boiling saturated aliphatic compounds occur naturally in aquatic organisms and probably are not toxic, but they may have detrimental effects. Chemical cues are used extensively by aquatic organisms for finding food, avoiding predators, locating mates, and navigating (e.g., the homing salmon finding its way back to the stream in which it was spawned). The concentration of indicator molecules detected by such animals is in the parts-per-billion range. Although only a few such indicator compounds have been identified, many are thought to belong to the high boiling saturated hydrocarbon group. Therefore the presence of larger quantities of related compounds as a result of an oil spill could disrupt chemical communication completely, either by mimicking the natural indicator compounds or by overloading the sense organs. The organisms affected would be unable to locate food, avoid predators, etc. This could have disastrous consequences for the populations involved. Such problems have not been conclusively demonstrated as yet, partly because the technical difficulties involved in demonstrating them are very great. The olefinic hydrocarbons, also produced by aquatic organisms, may be involved in chemical communication, although their biological roles are not well known.

In recent years much research effort has been applied to the subject of what to do about marine oil spills. Most projects have been concerned with methods for disposing of spilled oil, both before and after it reaches the coast. Much of the work has centered around devices for containment and recovery of floating oil slicks (Fig. 6-3). Methods for beach cleaning have been developed, as have techniques for dispersing the oil before it reaches the beach. Opinions are divided, but much of the available evidence indicates that the most desirable alternatives for dealing with oil spills are recovery or, if that is impossible, burning the oil with the aid of various "wicking agents." Burning does create a serious air pollution problem, but this is considered preferable to the damage done to aquatic organisms and the esthetic damage done to beaches. In spite of this high degree of research interest in oil spills, little work has been done on their effects on marine organisms. Most of the studies have concentrated on short-term effects on the large, mobile, and usually economically important species. It is well known that oil is bad for

Fig. 6-3. Massachusetts's DIP (Dynamic Inclined Plane) oil skimmer cleaning up an oil spill in Boston Harbor. The booms on either side help to direct the oil into the mouth of the skimmer. (Courtesy JBF Scientific Corp., Burlington, Mass.)

sea birds; it coats their feathers, rendering flying and sometimes swimming impossible. The weakened birds also become susceptible to pulmonary diseases like pneumonia. Even those that are rescued and cleaned have a 99% probability of death from the exposure to oil. Although the damage to bird populations is obvious, most of the studies on intertidal forms have indicated that they are relatively unaffected by crude oil spills.

The usual course of events after a crude oil spill follows this general pattern. Immediately after the spill the oil forms a floating raft on the surface of the water. Wave action tends to break up this mass of oil and disperse it as droplets in the upper layers of the water. The rate at which droplets are formed is a function of the severity of wave action, with relatively little breakup occurring in calm seas. Meanwhile, the oil is attacked by a host of decomposer organisms such as bacteria, molds, and yeasts. These organisms begin to break down the oil, attacking the smaller straight-chain saturated compounds first. Thus the decomposition process begins with the least toxic fractions of the oil mixture. It is a long process, requiring several months to convert the mass of oil into tarry lumps that still float on the surface of the open sea, and it is by no means certain that this bacterial breakdown leads to less harmful substances. The by-products of decomposition may be even more toxic than the original oil. In some parts of the ocean these lumps reach a concentration of 0.5 cm^2/m^2 of sea surface. Thor Heyerdahl encountered them all across the North Atlantic when he made his famous crossing in a papyrus boat in 1970. These lumps still contain some of the toxic, low boiling aromatic compounds after months at sea. They are frequently ingested by certain species of fish and in this manner become a part of the aquatic food chain. Meanwhile, the droplets of oil that have been separated from the main mass and are now suspended in the water may be ingested by filter-feeding organisms, thereby entering the food chain at another point. Some of this material also enters the bottom sediments, where detritus feeders may eat it, and this provides still another way in which the oil droplets can be introduced into the food chain. In the bottom sediments decomposition also takes place, but very slowly. While the unchanged oil droplets are in the bottom sediments, they can exert their toxic effects on the bottom organisms and larvae that attempt to settle on them. The presence of the oil in the sediment can alter its consistency enough to influence those organisms that are heavily dependent on sediment structure for their survival.

There is a lack of agreement among investigators as to whether the oil is ever broken down completely. Some say that decomposition of spilled oil is complete within a year, and others say that it may never occur. In many instances the oil has been considered to have been broken down simply because it is no longer visible, either on the surface of the sediments or in the water. This superficial kind of damage assessment ignores the oil trapped deep in bottom muck or tied up in organisms. Once the oil gains access to the system of a marine animal, it becomes bound in the fatty tissues where it may remain unchanged for years. Experimental evidence of petroleum fractions having been stored in animal fat unaltered for at least 2 years is now available. When held in this way, oil by-products can be accumulated in food chains in the same manner as pesticides (Fig. 1-11).

As mentioned earlier, refined petroleum products seem to be more toxic than crude oil. As an example, a small aviation gasoline spill off Santa Barbara in 1968 killed thousands of lobsters, fish, and beach invertebrates, whereas the thousands of tons of crude oil that leaked into the Santa Barbara channel in 1969 apparently had little effect on these species. The major study (directed by Dale Straughan) of the Santa Barbara incident has been criticized as not looking for the subtle effects or extending over a long enough time period to show them, but one would expect that kills of lobsters and fish would have been reported if they had occurred.

To date, the most complete before-and-after study of an oil spill is that of the *Tampico Maru* accident off the coast of Baja, California, in 1957. Diesel oil (a refined oil fraction) drifted into a cove, destroying all forms of marine life. A few hardy species had returned within 2 months, but it was 2 years before "significant improvements were noted." Four years after the spill, sea urchin and abalone populations were still greatly reduced, and 10 years later some species present before the spill still had not returned. The active agents in this case were probably the saturated low boiling hydrocarbons, which are the major components of diesel oil.

Another well-studied spill is that from the barge *Florida,* which ran aground off the Massachusetts coast in 1969, spilling 650 to 700 tons of No. 2 fuel oil. Here the course of the oil has been followed through the sediments and organisms. Toxic effects of the spill were still spreading through the sediments 8 months after the accident, in spite of the fact that casual observers had pronounced the affected area "as beautiful as ever" a few weeks after the barge went aground. By the time they made these pronouncements the animals killed during the initial spreading of the oil had had time to decay and disappear, but the oil trapped in the bottom sediment was still moving outward and killing more animals as it went. These new victims were on the bottom of the bay and not readily apparent to those individuals examining the shoreline for signs of damage. The researchers studying the effects of the spill had to establish new control plots farther and farther away from the initially damaged area as the oil continued to invade new territory. It is thought that the loss of the organisms in the sediment, coupled with the physical presence of the oil, increased the spread of the oil-laden material to previously uncontaminated areas. Two years after the spill the oil was decomposing, but traces of it still remained in the sediment.

Less research has been devoted to freshwater than to marine oil spills, but the consequences appear to be similar. Some freshwater species are particularly sensitive to such spills, whereas others are quite resistant. Crustaceans (crabs, shrimps, etc.) seem to be among the most sensitive to oil pollution, whereas certain species of annelids (worms) are among the most resistant, with fish falling somewhere in between. For fish the most toxic fractions of petroleum seem to be the naphthalenes (e.g., 1,2-dimethyl naphthalene, 1,6-dimethyl naphthalene, biphenyl). In addition, the products of bacterial breakdown of crude oil in fresh water are also toxic to fish.

Another possible serious consequence of oil spills is the potential for spilled petroleum products to accumulate other toxic substances. Many pesticides and toxic industrial chemicals such as the polychlorinated biphenyls are only slightly soluble in water but are quite soluble in oil. Spilled oil could absorb such materials from both air and water and accumulate them over considerable periods of time. Subsequent ingestion of the oil by an organism could then result in the administration of a high dose of the toxic material. With extensive areas of the sea's surface covered with oil and a constant influx of airborne pesticides, this possibility is a real one and should not be ignored.

Perspective. Oil pollution is destined to remain a serious problem for the foreseeable future. The increasing volume of oil shipped and the increasing size of the tankers carrying it give ample cause for concern. The construction of hot oil pipelines over permafrost in the Arctic is another possible source of oil spills. The lower temperatures found in the Arctic will materially impede natural breakdown processes, and as a result the effects of a spill occurring there will extend over a much longer period of time than they would in a warmer locale. Scientists still have a lot to learn about the specific long-term effects of petroleum products on organisms, but enough is known to class this problem as one of serious magnitude.

POLYCHLORINATED BIPHENYLS

Description, occurrence, and uses. The polychlorinated biphenyls (PCB's) are a group of complex organic compounds that have been used extensively for many years, but only recently have they been recognized as serious pollutants. PCB's are close chemical cousins of DDT and are of environmental concern for many of the same reasons (see below). Structurally, PCB's consist of a double ring with a variable number of chlorine atoms replacing hydrogen atoms at the sites indicated by x's. There are 210 theoretically possible PCB compounds, and probably at least a few molecules of each are present in the

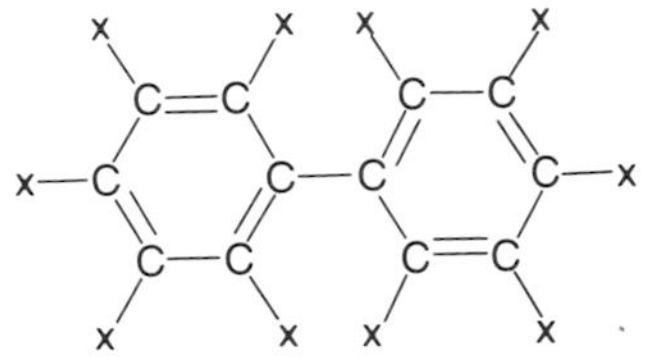

PCB (general formula)
x = Possible sites of chlorine substitution

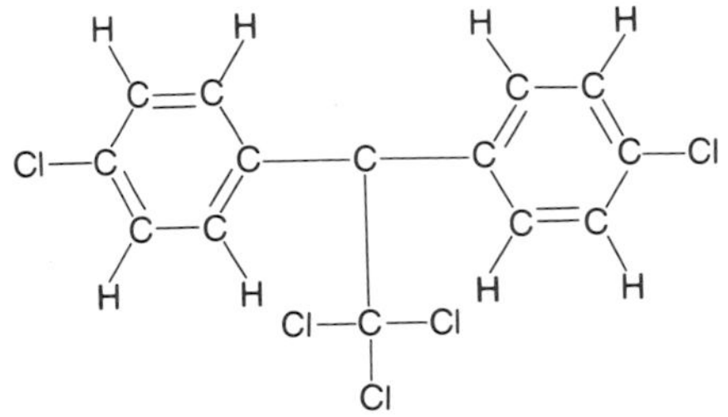

DDT(dichlorodiphenyl-trichloroethane)

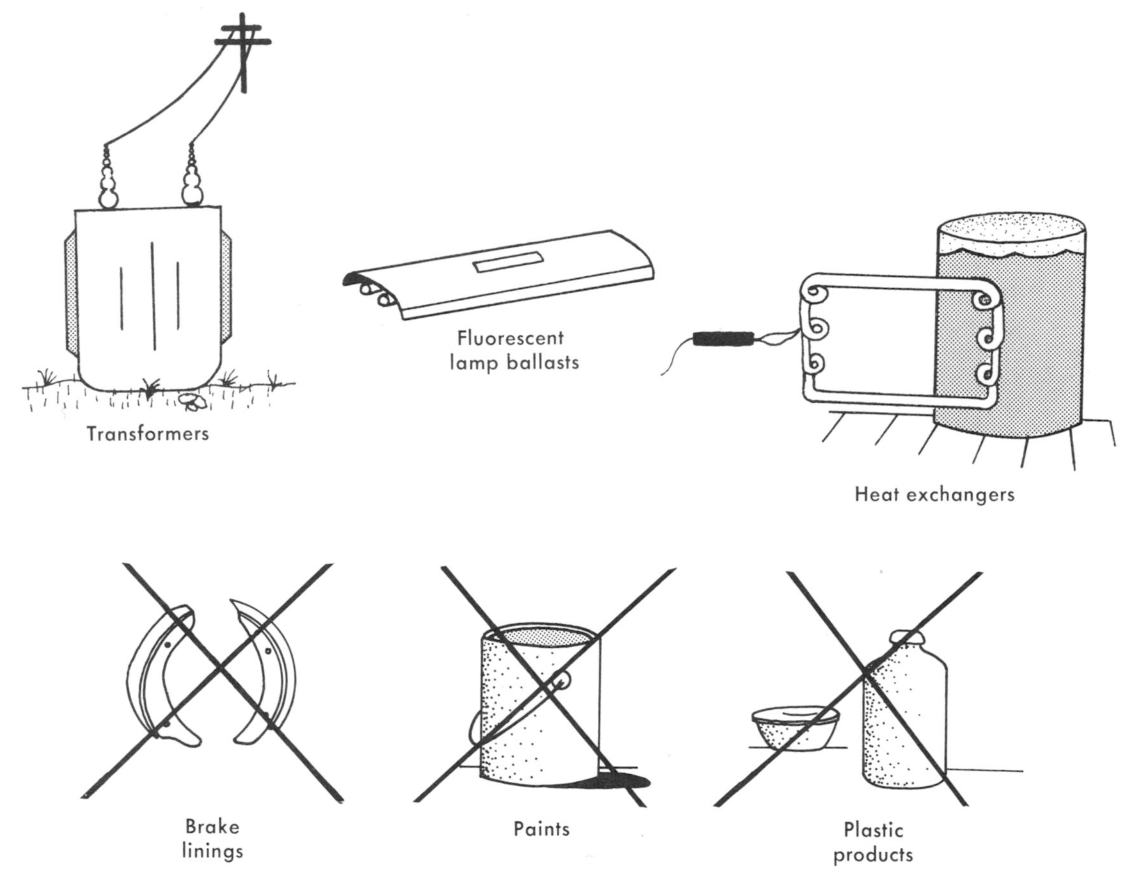

Fig. 6-4. Uses of PCB's. Those marked "X" have been eliminated as a result of the refusal of the only United States manufacturer, Monsanto Co., to sell PCB's for such applications.

currently available mixtures, although some forms are much more common than others. First reports of the occurrence of PCB's in the environment appeared in 1966 in samples from the Baltic Sea. Since that time they have been identified all over the world and have been implicated in human poisoning as well as causing damage to many kinds of organisms.

The biologically important characteristics of PCB compounds are their insolubility in water, their high solubility in fats, their toxicity to metabolic processes, and their stability (they are not readily attacked by any known organisms and do not break down rapidly when exposed to heat or sunlight). The low solubility in water coupled with a high solubility in fats provides a mechanism for accumulation in food chains (Fig. 1-11).

Polychlorinated biphenyls have had a great number of uses in recent years (Fig. 6-4). When they were first formulated, they were used as coolants and insulators in the large capacitors and transformers required by electric power companies—still one of the major uses. They are also used extensively as heat transfer fluids in industrial heat exchangers and as major constituents of hydraulic fluids. More recently they have been used as plasticizers in plastics and as additives in paints and in brake linings. They have been found to prolong and intensify the insecticidal powers of the chlorinated hydrocarbon pesticides and so have been incorporated as minor constituents of some insect killers.

Occurrence in the environment. Unlike DDT, most PCB's get into the environment accidentally rather than as a result of intentional spreading (Table 6-4). Some of the contamination almost certainly occurs during the manufacturing process, and additional amounts of PCB's are released in waste liquids and gases from plants using these compounds in their products. Leaks from heat exchangers are responsible for two of the most publicized recent PCB incidents, which were rice oil poisoning in Japan (see later) and chicken-feed contamination in North Carolina. Spills of PCB-containing hydraulic fluids are likely sources of some of the material found in the environment, and improper disposal of products containing PCB's undoubtedly account for the rest. Many of the objects burned or deposited in municipal incinerators and dumps (e.g., fluorescent light ballasts, oil-filled capacitors of discarded electronic equipment) contain significant quantities of PCB's. These are probably major sources of environmental contamination, either by release to the atmosphere through burning or by pollution of runoff and ground water from landfills. Most PCB release into the environment, from whatever source, is concentrated around highly industrialized areas.

TABLE 6-4. Sources of PCB's entering the environment in 1970 (North America)*

SOURCE	TONS
Disposal in dumps and landfills	18,000
Leaks and disposal of industrial fluids	4500
Vaporization of plasticizers	1500
Vaporization during open burning	400
Total	24,400

*Data from Nisbet, I. C. T., and Sarofim, A. F.: Rates and routes of transport of PCB's in the environment, Environmental Health Perspectives (experimental issue no. 1), pp. 21-38, April, 1972.

Occurrence in organisms. Concern for environmental contamination by PCB's first resulted from measurements of DDT levels in animals collected from the Baltic Sea. In those studies unidentified chemicals found by gas chromatography were traced to PCB compounds present in the samples. Further work has shown these compounds to be present in organisms of all trophic levels and in nearly all environments examined. The potential for biological concentration of PCB's through food chains has been mentioned previously, and the data of Table 6-5 seem to indicate that such concentration is occurring. Although the evidence for such concentration is not yet irrefutable, most investigators are convinced that it is going on. The concentration factor in food chains involving birds and mammals appears to be between 10 and 100 at each step, with the long food chains of marine ecosystems yielding PCB concentrations in the top predators that are about 1 million times higher than those in the seawater. Some measurements in marine birds indicate concentration factors of 10 to 100 million times more than those of the seawater.

Effects on organisms. As with most other forms of environmental pollutants, PCB's have both acute and chronic effects. When these compounds were first synthesized and came into wide industrial use, their acute toxicity was investigated and found to be quite low. The acute toxicity varies with the degree of chlorination of the compound (percentage of chlorine), and in all tests to date acute effects are produced at levels at least ten times (in some cases thousands of times) higher than those at which DDT produces similar effects. This low acute toxicity leads to a low level of concern for the fate of PCB

TABLE 6-5. Concentration of PCB's in samples from the Irish Sea and Firth of Clyde, Scotland*

SAMPLE	NUMBER OF SAMPLES	PCB (ppm)
Seawater	53	10 ppt (parts per trillion)
Zooplankton	7	0.01 - 0.03
Mussels	200+	0.05 - 0.5
Norway lobster	33	0.01 - 0.1
Herring	154	0.01 - 2.0
Whiting		
Muscle	18	0.01 - 0.4
Liver	15	1.0 - 7.0
Cod		
Muscle	5	0.3 - 1.8
Liver	5	4.5 - 50.0
Other fish	83	0.1 - 1.0
Razorbill (bird): livers	7	2.0 - 44.0
Guillemot (bird): livers		
Birds found dead	49	2.0 - 880.0
Birds shot	5	0.0 - 2.0

*Data from Holdgate, M. W. In Nisbet, I. C. T., and Sarofim, A. F.: Rates and routes of transport of PCB's in the environment, Environmental Health Perspectives (experimental issue no. 1), pp. 21-38, April, 1972.

compounds after they have served their useful life. The relatively few cases of human PCB poisoning have been attributed to chronic exposures, usually persisting for weeks or months, rather than acute exposures.

The chronic effects of PCB's on organisms are much more pronounced than the acute effects. This is to be expected when a material is concentrated in the fatty tissues of the organism—each exposure adds to the total already present in the body. The most notable case of human PCB poisoning occurred in Japan in 1969. About 1000 individuals are known to have become ill as a result of eating rice oil that had become contaminated by PCB's from a leaking heat exchanger. The average PCB dose of those affected was about 2 grams, whereas the lowest effective dose seemed to be about 0.5 gram. The effects extended to children born of mothers who ate the contaminated oil. The symptoms of human PCB poisoning include nausea, vomiting, jaundice, edema, abdominal pain, and chloracne (dark brown stains on nails, skin, and gums and pimples on the skin). The liver may be damaged severely by high doses of PCB's, leading to acute yellow atrophy and, ultimately to death if enough liver tissue is destroyed. Recovery is uncertain (victims of the Japanese incident still exhibited symptoms 3 years later), and there is no known treatment for PCB poisoning.

In birds the effects of chronic exposure to PCB's include widespread edema, enlargement of the heart, liver, and kidneys, and atrophy of the spleen. Decreased egg production, reduced egg hatchability, and an increase in the number of embryonic and post-embryonic deaths have also been reported. PCB's apparently do not, however, cause eggshell thinning in the way DDT and some other chlorinated hydrocarbon pesticides have been shown to do (Chapter 7). Fish also appear to be sensitive to PCB's. The observed toxicity does not seem to be as great as that of DDT (by a factor of about 100), but fish kills in the vicinity of PCB manufacturing plants have been traced to these chemicals. Shrimp and shellfish are also subject to PCB poisoning, with as little as 1 part per billion (ppb) constituting a lethal dose for some species of shrimp. By comparison, the most sensitive mammal tested to date is the mink, with a lethal level of 10 ppm. Some aquatic plants are also sensitive to PCB's, and it is at this point that most environmental concern over PCB's has been focused. Photosynthesis and growth have been inhibited in several species of marine diatoms by concentrations of PCB's of 10 to 100 ppb. If these organisms at the base of most marine food chains are seriously affected, this will eventually affect the availability of food from the sea.

Perspective. PCB's were originally used entirely in "closed system" applications, and there was little concern for the possibility of their escape into the environment. On discovering the scope of their dispersal in the world ecosystem, however, increasing care has been taken by the only United States manufacturer (Monsanto Co.) to ensure that escape into the environment will be sharply curtailed. The company has ceased to sell PCB formulations for any uses other than those of the closed system type. Suspension of sales to manufacturers of hydraulic fluid, plastics, paint, and carbonless carbon paper has reduced the total domestic sales volume of PCB's by about 40%. Monsanto Co. has also set up a PCB-disposal facility for safely destroying spent PCB compounds. The U. S. Food and Drug Administration has established a monitoring program for PCB levels in foods. Large numbers of PCB-contaminated chickens have been destroyed as a result of this program. The Environmental Protection Agency has set a standard of 0.01 ppb for PCB's in streams, and this level appears to provide an adequate measure of safety. In summary, the PCB's seem to constitute a potential source of environmental difficulty, but with worldwide application of the steps being taken in the United States the problems associated with potentially high levels should not materialize.

PHTHALATES

Description, occurrence, and uses. Phthalates are a group of chemical compounds (esters of phthalic acid) that have been used extensively as plasticizers in plastics, particularly those of the polyvinyl chloride (PVC) type. They are chemically related to thalidomide, a sedative drug that was responsible for a variety of birth defects in children born between 1958 and 1961 to women that had taken it during the early stages of pregnancy. Since the plasticizer may form up to 40% of the final plastic item, total use of these compounds is very great. In 1968 United States production of phthalates totaled approximately 840 million pounds.

In the manufacture of plastic, molecules of a resin or monomer are linked together, forming long chains of polymers. The plasticizer is then added to give the final product flexibility. The plasticizer molecules do not form strong bonds with the polymer molecules but, rather, occupy the spaces between them. This provides a "lubricating" effect, allowing the polymer molecules to slide past one another. Because the plasticizer molecules are not bound firmly, they can escape from the plastic, either into the atmosphere or into liquids with which they come into contact. This mobility of the plasticizer molecules is responsible for their presence in the oily film found on the interior of automobile windows and also for the high concentrations measured in PVC-packaged human blood. The "new-car smell" (Fig. 6-5) has been attributed to these same phthalate compounds, and it is in the interiors of new automobiles that most of us are exposed to the highest concentrations of them. Probably more serious, however, is the massive dose of phthalates received by critically ill hospital patients who require large quantities of transfused blood. Concentrations of the plasticizer di(2-ethylhexyl)phthalate (DEHP) of 5 to 7 mg per 100 ml have been measured in whole blood stored in plastic containers for the maximum allowable period of 21 days. Thus a patient needing a complete replacement

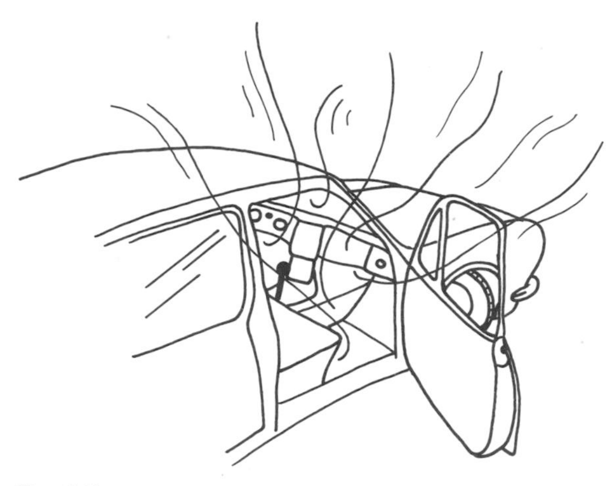

Fig. 6-5. The "new-car smell," familiar to car buyers for a generation, is attributable to the phthalate compounds released from the many plastic materials used in the car's interior.

transfusion of 14 pints could receive a dose of 450 mg of DEHP. Introduction of this quantity of any foreign chemical might be expected to cause difficulty in normal individuals, let alone those who are already critically ill.

Effects on organisms. The phthalates used in plastics have been tested in laboratory animals and found to be only slightly toxic when administered orally or injected into the body cavity. Lethal doses for mice are in the range of 1 to 10 grams per kilogram of body weight, a very low order of toxicity. However, recent studies have demonstrated toxicity to embryos at very much lower doses. In these experiments the LD_{50} (dose at which 50% of the test animals die) of eight phthalates were determined for adult female rats. Dosages of 1/10, 1/5, and 1/3 of the LD_{50} for each compound were administered to pregnant rats on the fifth, tenth, and fifteenth days of gestation. Both compound-dependent and dose-dependent effects were observed in the resultant offspring. Effects ranged from reduced fetal weight to complete resorption of the embryos. Many kinds of abnormalities (e.g., absence of tail or eyes, twisted hind legs, fused ribs, and abnormal skull bones) were encountered. DEHP proved to be one of the least toxic of the phthalates tested, but it caused about 27% resorption of embryos, and of the remainder, 22% were grossly abnormal.

The mechanisms whereby the phthalates produce their effects on mammalian systems are largely unknown. It is known that butyl glycolylbutyl phthalate is metabolized to glycolyl phthalate by rat livers. However, in the same experiment DEHP was not metabolized but was stored in the liver. DEHP has been identified in the spleen, liver, lungs, and abdominal fat of human patients who had received transfusions of whole blood from PVC bags. The highest concentrations were found in samples of abdominal fat. Although the concentrations observed in humans were well below those found to cause embryonic abnormalities in rats, the tendency of DEHP to accumulate in the tissues could lead to a buildup of dangerous proportions in individuals requiring repeated massive transfusions. In another report hepatitis-like symptoms in a group of artificial kidney patients were traced to the 10 to 20 mg per liter of diethyl phthalate leached from PVC tubing contained in a new dialysis machine. No cause-and-effect relationship was proved, but when the patients were transferred to an older machine without the PVC tubing, the symptoms abated dramatically.

Perspective. The vast quantities of phthalate-containing plastics presently in use and being discarded provide large sources for these potentially dangerous compounds. Luckily, they are not toxic in low concentrations, but they are capable of accumulating, both within food chains and in individual organisms. More research needs to be done on the mechanisms by which they cause their toxic effects, and more attention to their use in medical and surgical applications is called for. The phthalate plasticizers provide another example of a class of materials that have been considered biologically inert, but are not.

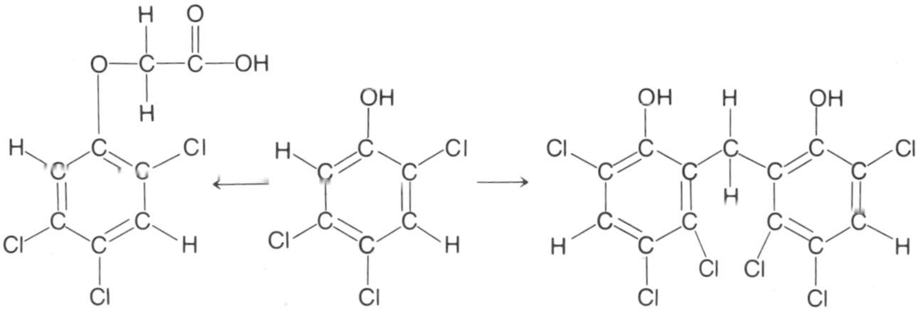

2,4,5-T — 2,4,5-Trichlorophenol — Hexachlorophene

HEXACHLOROPHENE

Description and uses. Hexachlorophene is an ingredient in many soaps, disinfectants, and cosmetics. It is used because of its bactericidal properties, and it has become synonymous with cleanliness in the minds of many Americans. Chemically, hexachlorophene is related to the herbicide 2,4,5-trichlorophenoxyacetic acid (2,4,5-T), and it is prepared from the same precursor compound, 2,4,5-trichlorophenol. 2,4,5-Trichlorophenol is known to give rise to dioxins, materials thought to be responsible for the birth defects (in animals) that had originally been attributed to 2,4,5-T. Dioxins show toxic effects at concentrations below current limits of detection so it is possible that they could also be responsible for the toxic effects ascribed to hexachlorophene. The toxic properties of hexachlorophene have been known since shortly after its first synthesis, but until recently, little attention had been given to the potential dangers in its excessive use. The cases of chloasma (a darkening of the face) and burn encephalopathy (brain damage) that were reported during the 1960's did not trigger a new investigation of the safety of hexachlorophene in spite of the association, in each case, between the symptoms and heavy use of a 3% solution of the material. Chloasma resulted from the use of such solutions in the treatment of severe acne, and burn encephalopathy occurred in burn patients whose burns were cleaned frequently with hexachlorophene to prevent staphylococcal infections. New toxicity studies were initiated by the U. S. Food and Drug Administration in 1969 when a manufacturer applied for a license to use hexachlorophene in a fungicide preparation. The results of those studies clearly implicated hexachlorophene as the causative agent in several diseases in animals.

Effects on organisms. Hexachlorophene has been found to produce brain and spinal cord lesions leading to paralysis in mice. These lesions appear as an edema (swelling) of the white matter of the brain and spinal cord, and they are reversible if the animal is removed from exposure to hexachlorophene. In an attempt to relate these findings to human exposures the concentration of hexachlorophene in the blood of affected rats was measured and found to average 1.2 ppm. Checking a sample of twelve human subjects with no particular hexachlorophene exposure revealed levels ranging from 0.005 to 0.89 ppm in their blood. In another study the blood of infants being discharged from a hospital, where bathing babies with a strong hexachlorophene solution was routine, was checked for hexachlorophene and found to contain an average of 0.109 ppm with a maximum of 0.646 ppm in one child. The latter level is more than half of that which has been found to cause brain lesions in rats. Another study in which baby monkeys were washed with a 3% hexachlorophene solution every day for 90 days demonstrated brain damage similar to that observed in rats. These findings have resulted in actions by the U. S. Food and Drug Administration designed to limit accidental exposure to hexachlorophene. It will no longer be allowed as a significant ingredient in cosmetics (e.g., feminine hygiene sprays), but it will remain available for use against staphylococci in specific situations.

Perspective. The case of hexachlorophene provides an object lesson in the handling of a useful but toxic material. On recognition of its dangerous properties, hexachlorophene has been placed in the category of drug, and its use will be regulated in the manner of such substances. With the implementation of this action hexachlorophene should cease to pose a serious hazard to human health.

Note: Since this was written, the hexachlorophene-caused deaths of twenty-four French infants has been reported. The babies were dusted with a talcum powder that contained 6% hexachlorophene, and all died of apparent severe brain damage. Since this announcement, French authorities seized all unsold stocks of the powder and issued warnings to mothers not to use any that they might still have.

SUGGESTED READING

Combustion hydrocarbons

Olsen, D., and Haynes, J. F.: Preliminary air pollution survey of organic carcinogens—a literature review, Raleigh, N. C., 1969, U. S. Department of Health, Education and Welfare.

Stern, A. C.: Air pollution, ed. 2, vol. I, New York, 1968, Academic Press, Inc.

Solvents

Bowen, D. H. M.: Solvent vapors under fire, Environmental Science and Technology **5**:1086-1087, 1971.

Oil

Blumer, M., Sanders, H. S., Grassle, J. F., and Hampson, G. R.: A small oil spill, Environment **13**(2):2-12, 1971.

Smith, J. E., editor: *Torrey Canyon* pollution and marine life, London, 1968, Cambridge University Press.

Polychlorinated biphenyls

National Institute of Environmental Health Sciences: Environmental Health Perspectives, April, 1972. (First experimental issue devoted entirely to papers on polychlorinated biphenyls.)

Peakall, D. B., and Lincer, J. L.: Polychlorinated biphenyls—another long-life widespread chemical in the environment, Bioscience **20**:958-964, 1970.

Phthalates

Jaeger, R. J., and Rubin, R. J.: Plasticizers from plastic devices: Extraction, metabolism, and accumulation by biological systems, Science **170**:460-462, 1970.

Singh, A. R., Lawrence, W. H., and Autian, J.: Teratogenicity of phthalate esters in rats, Journal of Pharmaceutical Science **61**:51-55, 1972.

Hexachlorophene

Kimbrough, R. D., and Gaines, T. B.: Hexachlorophene effects on the rat brain, Archives of Environmental Health **23**:114-118, 1971.

Wade, N.: Hexachlorophene: FDA temporizes on brain-damaging chemical, Science **174**:805-807, 1971.

7 AGRICULTURAL ORGANIC COMPOUNDS (PESTICIDES)

Perhaps the most relentless battle ever waged by man in all his history has been against insects and other pests that would destroy his crops. The harmful insects (Fig. 7-1) probably account for no more than 0.1% of the known species, but this fraction has ravaged man's agriculture for all of recorded history. There is no doubt that mankind's population has been held in check by his limited food supply, and insect pests have been a prime factor in this limitation.

Some liberation came for man when he began using certain heavy metals such as lead and the metal-like arsenic to poison and so reduce the numbers of these pests. Although these poisons are still in use today, the revolution in pest control—the breakthrough—came during World War II with the introduction of synthetic, organic pesticides, notably DDT. The synthesis of several other biologically active pesticides was soon accomplished. For a while insect populations dwindled and crop production soared. The possibility of feeding a hungry world was in sight. In 1945 United States production of DDT was 33.2 million pounds. By 1967 the annual production of a variety of pesticides reached over 1 billion pounds, valued at about 900 million dollars. Production and value of pesticides has increased steadily. Over 3 billion pounds of DDT alone have been produced in the United States since 1944.

As the use of DDT to control crop-destroying insects increased, so did the insects' resistance to it. This stimulated the search for and synthesis of newer, more toxic chemicals. To date something in excess of 900 synthetic chemicals incorporated into some 45,000 registered pesticides are or have been on the market. These find use in small and large agricultural operations, in forestry work, in urban sanitation, in food storage, and in almost every home, garden, and lawn in the nation.

The pesticides can be divided into the following four major categories according to their use:

1. *Insecticides.* These chemicals, as the name implies, are intended for the control of insects.
2. *Herbicides.* These chemicals kill undesirable weeds and shrubs. They are defoliants and desiccants.
3. *Fungicides.* These compounds destroy undesirable molds (fungi).
4. *Other specific pesticides.* These include rodenticides, molluscicides (against shellfish), nematocides (against roundworms), algicides (against algae), and a variety of repellents and attractants designed to be effective against ants, mites, ticks, and other animal forms.

Although the chemicals in each of these categories were designed to be lethal to only certain animal forms, they have proved to possess a considerably wider range of toxicity. Some of the insecticides, for example, in addition to being effective against the primary target form are also lethal to fishes, birds, and small mammals. A truly specific pesticide, lethal

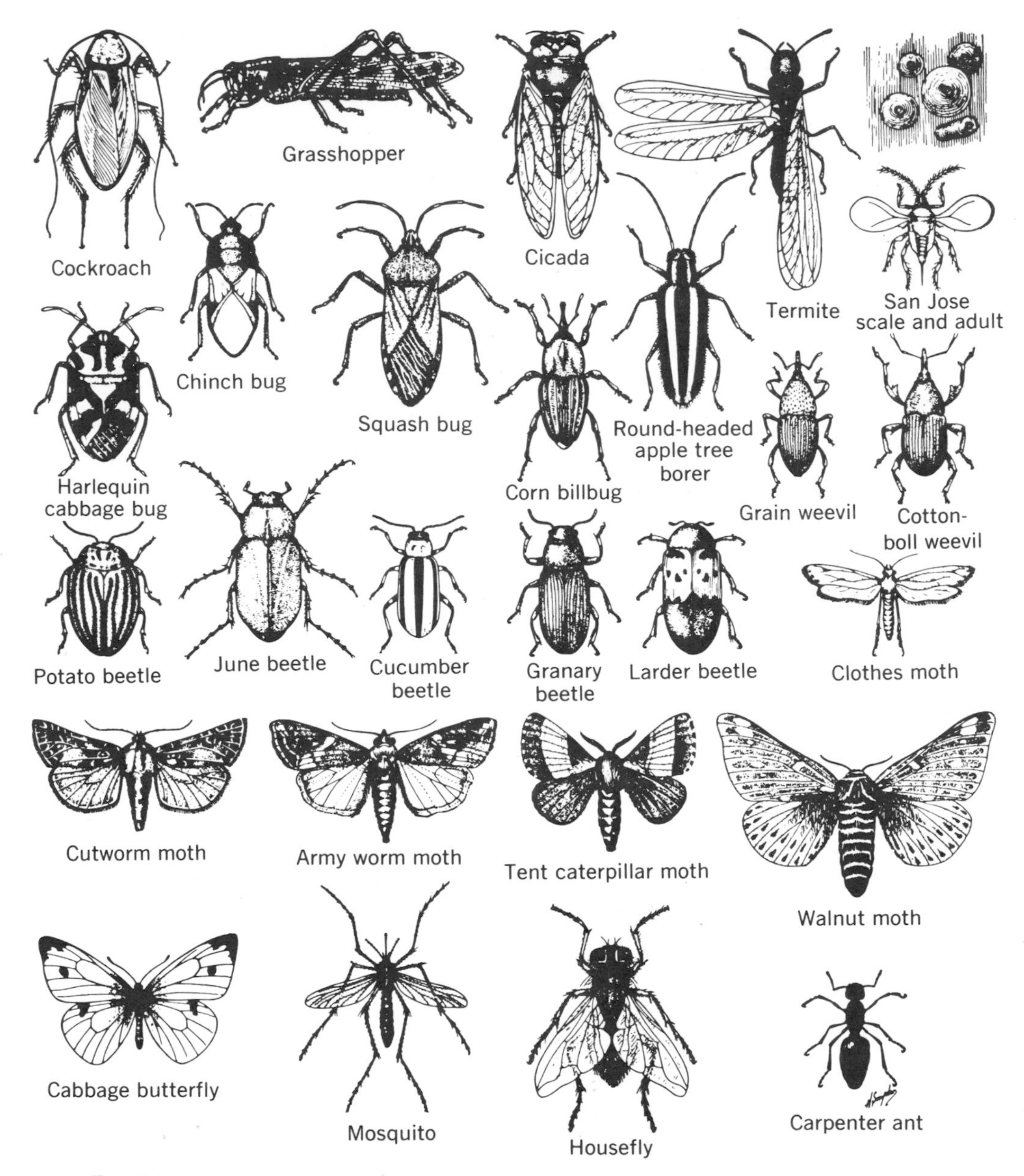

Fig. 7-1. Some harmful species of insects against which insecticides are designed to act. Harmful insects represent only about 0.1% of all insect species. (Courtesy General Biological Supply House, Inc., Chicago, Ill.; from Beaver, W. K., and Noland, G. B.: General biology, ed. 8, St. Louis, 1970, The C. V. Mosby Co.)

only to one type of organism, has proved to be difficult to design as well as being economically less desirable than compounds that are effective against a variety of pests.

From a pollution point of view the main problem with many pesticides, especially with certain kinds of insecticides such as the chlorinated hydrocarbons (see later), is their long persistence in the environment. DDT and related insecticides are not ordinarily metabolized (broken down) by living or nonliving systems. They are nearly insoluble in water but are soluble in fats so that they tend to accumulate in the fatty tissues of living organisms. They are mobile in the environment in that these chemicals can bind to airborne particulate matter and be distributed globally or transported in suspensions by river and ocean currents. Much of the 3 billion pounds of DDT that have been manufactured is still present as DDT or closely related chemical compounds. It has been found in every part of the earth and its waters that have been examined for it.

The fact that several of the chlorinated hydrocarbons are nerve poisons, upset normal functioning of enzyme systems, exhibit hormonelike activity, inhibit photosynthesis by plants, and are carcinogenic in test animals makes the extreme health hazard potential of this class of pesticides obvious. Some of the more common pesticides will be examined according to their use, environmental persistence, toxicity, etc. Two types of pesticides, insecticides and herbicides, will be treated in some detail.

INSECTICIDES

All of the pesticides discussed in this chapter belong to the newer classes of man-made chemical compounds. They have largely replaced the metallic and metal-like compounds (arsenates, mercurics, etc.) and the natural plant products (pyrethrins, rotenone, nicotine) that had served as insecticides in earlier times. The newer chemicals were at once cheaper to make and more lethal to the target insects, and so they have found favor with the agricultural community that uses the bulk of these products. The insecticides currently in use may be grouped into three categories according to their chemical structure. These are the chlorinated hydrocarbons, the organophosphates, and the carbamates.

Chlorinated hydrocarbons

The chlorinated hydrocarbons are composed of carbon and hydrogen, to which are attached chlorine atoms. This family of compounds includes DDT, dieldrin, lindane, aldrin, endrin, heptachlor, chlordane, toxaphene, and several others. Of these, DDT (dichlorodiphenyl-trichloroethane) has been studied most extensively. It has been used more widely and so is more widespread in the environment than any other pesticide. Originally designed to kill harmful insects, its broad spectrum of effects now includes other arthropods (shrimps, crabs, lobsters), shellfish, worms, fishes, reptiles, birds, and mammals. Since a major effect of DDT is on the nervous system, *any* animal may be subject to its lethal effects and those of its chlorinated hydrocarbon relatives. At less-than-lethal doses DDT causes increased activity due to overstimulation of the nervous system and various unusual behavior patterns.

The toxicity of the chlorinated hydrocarbons is compounded by their long persistence in the environment. All are synthetic (man-made) compounds, and organisms generally do not have mechanisms for breaking them down or detoxifying them. They therefore accumulate when introduced into the environment. They are estimated to have a half-life of up to 15 years. When metabolic breakdown occurs, as it can in the liver very slowly, the principal product is DDE (dichlorodiphenyl-dichlorethylene). This product is still toxic, however, so little is gained by metabolizing DDT. Most of the environmental load of the chlorinated hydrocarbons is believed to be in the form of DDE. DDE and DDT are almost insoluble in water but quite soluble in organic solvents (e.g.,

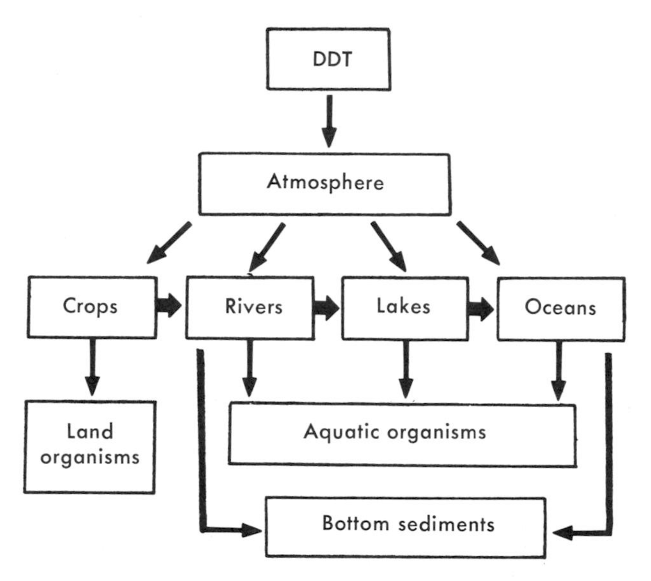

Fig. 7-2. Pathways taken by DDT after it is added to the environment by man. As shown, DDT does not remain at the site of application. It is estimated that only one half of the DDT sprayed on an agricultural crop actually reaches the target plants. The remainder is transported to aquatic environments and other organisms.

benzene) and in fat. This latter fact means that animals can "attract" and concentrate the chlorinated hydrocarbons from a watery environment.

DDT and other chlorinated hydrocarbons do not remain in toto on their site of application (Fig. 7-2). When applied as aerosols or sprays, much of the insecticide will be picked up by wind and dispersed. It is estimated that less than half of the DDT sprayed by crop-dusting aircraft ever reaches the ground. Much of the DDT sprayed on an area is carried off with surface water, but more is lost to the atmosphere by being adsorbed to particulate matter. DDT is also carried by rivers to the lakes and oceans and, of course, is transported by animals when they concentrate it. Air transport of the chlorinated hydrocarbons is especially fast; they can be distributed around the world in a few weeks from some sites of application.

The greatest danger to living animals from the chlorinated hydrocarbons lies in their ability to concentrate these insecticides. An organism at one trophic level consumes many organisms from the level just below it (Fig. 1-11). One bird eats many worms, for example. Although each worm may have only small quantities of an insecticide in its tissues, the bird, in a lifetime of metabolizing earthworms, will accumulate increasing amounts of the poison each time it eats. The insecticide cannot be detoxified by the bird, and it accumulates in its fatty tissues. The insecticide may be in the air or water in insignificant amounts but could become 1000 times more concentrated in the predator animal than in the animal preyed upon, and many times more than in the atmosphere or waters.

There is no question that the DDT family of insecticides has provided mankind with great benefits in the form of increased crop yields, but the high toxicity and its indiscriminate effects on all forms of animal life have required us to take a hard look at the worth of its effects on destructive pests. For example, large tracts of evergreen forests in Canada have been sprayed with DDT in an effort to control the spruce budworm. In one case the trees around the watershed of a river system, normally supporting abundant fish populations, were sprayed with ½ pound of DDT per acre. In that year (1954) no salmon fry were seen, indicating a complete reproductive failure. Populations of older fishes were reduced by 50%, and in some cases the population of less-than-year-old fish was reduced to under 10% of that in unsprayed streams. These lethal effects extended over 30 miles downstream from the area sprayed. With the coming of winter many of the survivors, already weakened, succumbed. For some years later, fewer and fewer adult salmon were caught, reflecting the reproductive failure of 1954. Various other animals were also affected by this insecticide program. Aquatic insects (food of salmon) were decimated. DDT accumulated in earthworms, which formed the staple diet of at least one bird species (American woodcock). The birds and their eggs became contaminated, and breeding success was sharply curtailed. Numerous reports show that DDT interferes with eggshell manufacture, and infected birds lay thin-shelled eggs. Such malformed eggs are subject to easy breakage, jeopardizing reproductive success.

Carnivorous (meat-eating) birds are near the top of their food chains and, consequently, they receive high concentrations of the chlorinated hydrocarbons along with their food. A bird census in Connecticut in 1938 showed the osprey population to consist of about 200 pairs. By 1965 the number was reduced to 12 pairs. Productivity had dropped from an average of about 2.5 young per nest to 0.5 young per nest. The eggs contained 5 ppm of DDT, an amount sufficient to account for the decimation of the population over the 20-year period since the introduction of DDT. Ospreys are fish eaters, and they accumulated DDT from this source. The use of chlorinated hydrocarbons in England in the early 1950's resulted in a sharp population decline of a species of falcon. The same species was wiped out in the 1950's in the eastern United States. In both populations abnormal behavior was noted: breakage and eating of thier own eggs, abandoning the nest, etc. Tissues of the birds and their eggs contained DDE, heptachlor, and dieldrin.

The Lake Michigan aquatic ecosystem is apparently completely permeated with DDT (Fig. 7-3). The bottom mud contains about 0.014 ppm of DDT, some small arthropods contain thirty times more (0.41 ppm), the fishes contain ten times more than the arthropods (3 to 6 ppm), and herring gulls contain over twenty times more than that in the fishes (99 ppm). Breeding success in the gulls was poor, and they exhibited behavioral abnormalities. Another phenomenon was observed that appears to be general in nature. Starved birds developed nervous system anomalies and died, whereas well-fed birds survived, although carrying the same DDT concentra-

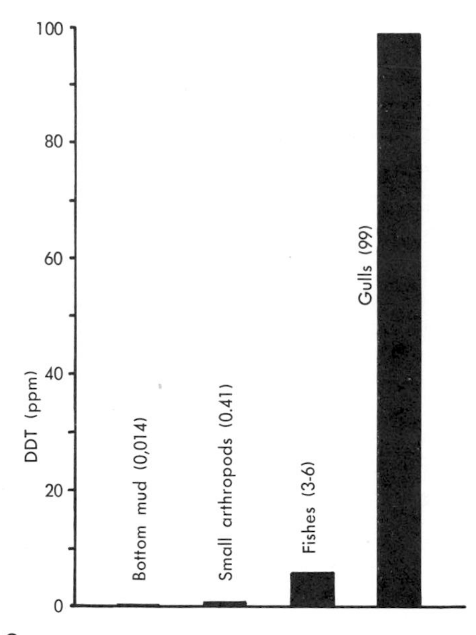

Fig. 7-3. DDT load in Lake Michigan and its concentration in bottom sediments and in an aquatic food chain. The DDT load in gulls in about 240 times that in the small arthropods.

tions. The reasons for the lethal effects of DDT on starved birds (and presumably on all other animals) is straightforward: A starving animal draws on its fat stores for energy; the DDT accumulated in these stores is thus released to the general circulation, where it can attack the nervous system and eventually cause death.

The earth's water cycle (Fig. 1-7) requires that all surface waters eventually will find their way to the oceans. These waters will carry with them much dissolved or suspended material from terrestrial ecosystems. In this manner much of the world's load of suspended chlorinated hydrocarbons, plus that in the atmosphere, will eventually be deposited in the seas. Research reports indicate that this is precisely what is happening. Oceanic birds (the Bermuda petrel, for example) that have had no contact with any large land mass that has undergone insecticide treatment, are exhibiting reproductive decline. Petrel chicks are carrying an average load of over 6 ppm of DDT. Some marine fishes are carrying heavier loads than freshwater fishes from lakes that have been exposed to a DDT-spraying program.

At the other end of the food chain, experiments show that small amounts of DDT in water suppresses photosynthesis of some marine algal forms. This effect is potentially very serious, since considerably more than half (some authors say 90%) of all photosynthesis takes place in the oceans. These algal cells are at the very bottom of marine food chains, and any disruption or imbalance at this trophic level will have far-reaching effects on every trophic level above it. If some undesirable species of algae were less susceptible to DDT and replaced a more susceptible food species, an entire food chain from top to bottom could collapse. Many organisms are extraordinarily sensitive to chlorinated hydrocarbons. Some shrimplike animals and some shellfish are sensitive to DDT in the parts per billion range!

Several authors have clearly detailed the "simplifying" effect of insecticides on world ecosystems; that is, insecticide use reduces the diversity of species in the treated area, and often well beyond the treated area. Natural catastrophes like fire accomplish the same thing. Such "simplifiers" favor survival of the smaller, more rapidly reproducing species near the bottom of food chains. It has been pointed out that large populations of small animals enjoy a greater genetic diversity than the generally smaller populations of large animals. In a large population there will always be certain individuals that will be resistant to pesticides, and these will be the members that survive an exposure to spraying, and they will be able to replenish the population. Large animals with their smaller total numbers may be completely eliminated, thus "simplifying" the ecosystem. Such selection of resistant individuals for survival is apparently the reason why several strains of the malaria mosquito and housefly have become virtually immune to DDT, despite very heavy applications of the insecticide.

Mechanism of action. Although the precise biochemical action of the chlorinated hydrocarbons is unknown, they are known to attack the central nervous system. One effect of DDT and its chemical relatives is to stimulate the nervous system, causing hyperactivity and muscle tremors (Fig. 7-4).

A second effect of the chlorinated hydrocarbons is on hepatic enzyme induction, that is, stimulation of enzyme synthesis. The enzymes induced are relatively nonspecific, but one result of increase in their synthesis is a reduction in the quantity of certain hormones (e.g., estrogen) in the blood. Such reductions have been noted in birds and in man. In birds such hormone imbalance results in the behavioral and reproductive changes already noted.

A third documented effect of the chlorinated hydrocarbons is the result of their inherent carcinogenicity. DDT has increased the rate of tumors in mice, rats, and trout, whereas aldrin, dieldrin, heptachlor, Strobane, and Mirex are carcinogenic in mice. Since human experimentation is not practical, no positive carcinogenic effects of the chlorinated hydrocarbons have been reported. Some studies of patients

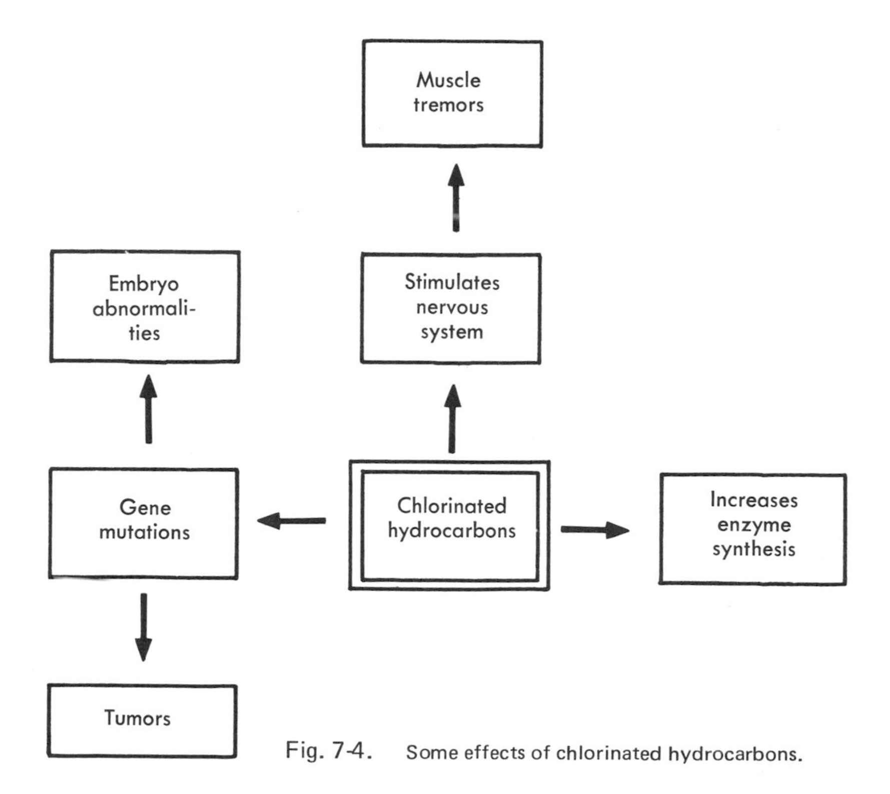

Fig. 7-4. Some effects of chlorinated hydrocarbons.

who have died from cancer show that these persons harbored substantially higher levels of DDT and dieldrin than those "normally" present in the general population. Although convincing human data are lacking, the results with experimental animals should be viewed with some alarm because there is often a high correlation in physiological reactions between mice and men.

A fourth effect of DDT is a genetic one. DDT has been shown to cause gene changes (mutation) in rats. Such mutations almost always alter the normal functioning of cells and in extreme cases can result in cancerous growth or teratogenic (abnormal development of the embryo) changes. Again, no specific cases involving mutagenesis by insecticides in man are known, but by inference the results observed in experimental animals warns of the possibility of similar effects in our species.

With the realization of the broad-spectrum toxicity and long persistence of the chlorinated hydrocarbons have come pressures from ecologically minded individuals and organizations to control the use of these pesticides. A number of court battles in the past several years have resulted in an almost total ban of the use of DDT in the United States (effective December 31, 1972). Environmentalists are currently attempting to make illegal the use of two other persistent chlorinated hydrocarbons: aldrin and dieldrin. It appears that, in time, most long-lasting insecticides may be banned for use in the United States, but because they break down so slowly, barring their use does not remove them from the environment. No one knows how much of the billions of pounds of the chlorinated hydrocarbons that have been used since 1945 are still in the environment—probably most of them.

Organophosphates

During the 1950's and 1960's, a group of complex organic insecticides containing phosphorus came into general use. These are the organophosphates. These compounds include parathion and methyl parathion, malathion, disulfoton, Guthion, ronnel, TEPP, Systox, and many others. The organophosphates are often more toxic (to man) than the chlorinated hydrocarbons but have the advantage (for the environment) of a short life: they are not persistent and so are effective only in the immediate area of their application. Some of the organophosphates, in fact, have a lower level of toxicity to vertebrates than DDT (e.g., malathion and ronnel). Certain others (TEPP) are, however, at least 100 times more toxic to rats than DDT. Because of their low persistence, the organophosphates may largely replace the chlorinated hydrocarbons as insecticides in the future.

The organophosphates appear to be, in general, much more toxic to insects than to man, but like the chlorinated hydrocarbons, they poison harmful and beneficial insects indiscriminately. These compounds, therefore, act like the DDT family in "simplifying" local ecosystems, that is, reducing species diversity.

Several cases of human poisoning by the organophosphates have been recorded. Physiological reaction to poisoning is immediate and involves nausea and headache, vomiting, and abdominal cramps. Heavy exposure may cause coma and death from respiratory failure in 5 minutes. Several such deaths from organophosphates have been recorded each year.

Mechanism of action. The biochemical action of the organophosphates has been determined (Fig. 7-5). These compounds inactivate the enzyme acetylchlolinesterase, which breaks down the nerve impulse transmitter acetylcholine. This effect allows acetylcholine to remain at the nerve synapse (Chapter 3), promoting a continued firing of impulses to muscles. In experimental animals this results in hyperactivity, uncontrolled twitching and tremors, and in extreme cases, death. Other enzymes are also inactivated by the organophosphates. Many of these insecticides, it was noted earlier, are more toxic to insects than to mammals, including man. One reason for this is the presence in mammals of an enzyme carboxyesterase, which destroys certain of the organophosphates (e.g., malathion). Some other organophosphates are known to inhibit this "protective" enzyme, however.

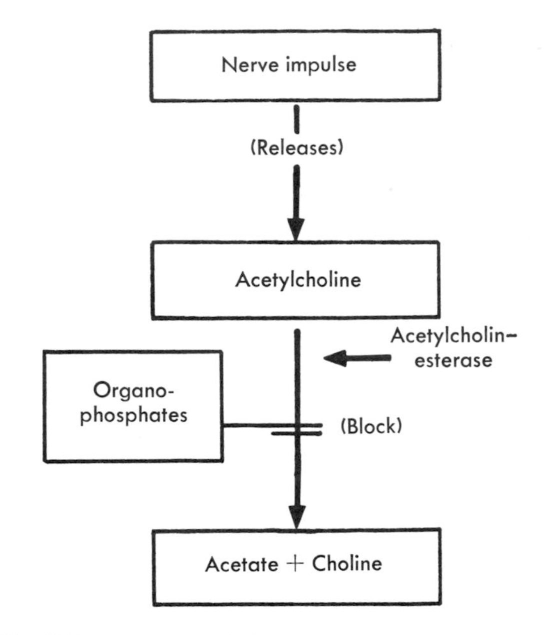

Fig. 7-5. Mechanism of action of the organophosphates. They block the action of the enzyme acetylcholinesterase at the nerve synapses. The result is a continued nerve stimulation resulting in muscle tremors and twitches.

Except for the inherent high human toxicity of some of the organophosphates, these insecticides are probably less dangerous to global ecosystems than the more persistent chlorinated hydrocarbon family. The effects of the organophosphates are limited to the local area of application much more than are those of the DDT family.

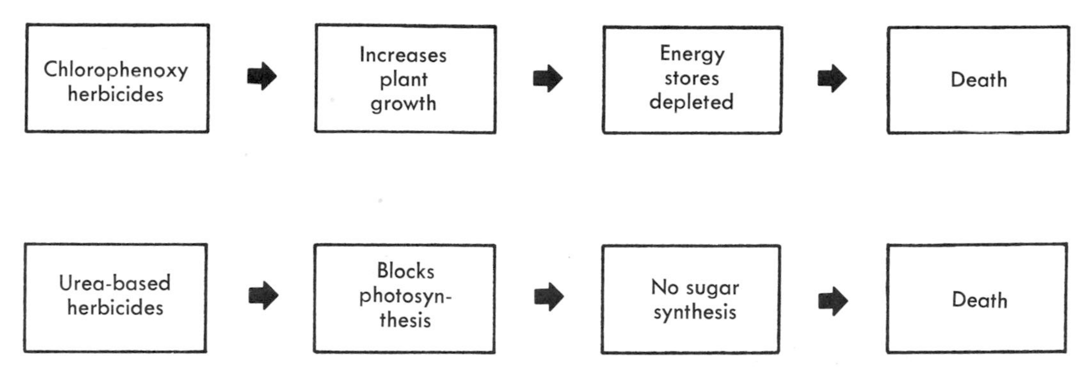

Fig. 7-6. Effects of the chlorophenoxy and urea-based herbicides on plants. The chlorophenoxy herbicides kill plants by causing more rapid than normal growth, depleting energy reserves. The urea-based herbicides block photosynthesis and so prevent growth.

Carbamates

The carbamates comprise a class of pesticides of more recent development. Members of this family are used as insecticides (Baygon, carbaryl), fungicides (nabam, zineb, ferbam), and molluscicides against snails and slugs (Zectran). The carbamates are members of a group of organic compounds called urethanes, derived from carbamic acid (H_2N–COOH).

The carbamates, like the organophosphates, are nonpersistent pesticides in that they are degraded rather rapidly to relatively nontoxic products. Also like the organophosphates their major effects are limited to the area of application. The toxicities of the carbamates vary, but generally they are less toxic than the organophosphates and considerably less so than the DDT family of chlorinated hydrocarbons.

HERBICIDES

The use of herbicides in the control of undesirable plants has increased steadily since World War II. In 1945, 917,000 pounds of 2,4-D was produced. Annual production by 1967 had exceeded 77 million pounds, and annual sales now exceed $25 million. Total production of all herbicides in the United States in 1967 was over 348 million pounds, having a sales value of about $430 million. The herbicides in that year made up 32% of all pesticides produced.

These compounds are used to destroy noxious weeds and shrubs that would otherwise overgrow roadsides, superhighway medians, railroads, power-line right-of-ways, and suburban lawns and gardens. The herbicides generally are more toxic to broad-leaf plants than to the grasses. This difference in action is mainly due to the ability of such broad-leaf species to absorb the herbicide more rapidly. These compounds also act on woody plants, but the treatments usually result in temporary defoliation rather than death. Repeated defoliation may, however, exhaust the tree's stored food reserves and thereby kill it.

In general, two classes of herbicides are in use. One, represented by 2,4-D, 2,4,5-T, and several others, are chlorophenoxy acids. The other group, represented by fenuron and diuron, are derivatives of urea (and also contain chlorine). The phenoxy herbicides act like certain natural plant growth hormones (auxins) in that they affect the growth rate of stems and roots and certain other aspects of plant physiology (Fig. 7-6). In the doses of phenoxy herbicides used plant growth is stimulated abnormally, and the plant literally grows itself to death by using up all its energy stores.

The urea-based herbicides have a quite different

action. These compounds interfere with a critical electron transfer reaction in photosynthesis. With photosynthesis shut down the plant cannot make its sugars or proteins and starves to death.

Certain newer compounds known as the triazines and the acylanilides are also in general use. These include Simazine and propanil, the former a triazine, the latter an anilide. Propanil has been used to kill barnyard grass in rice fields, dramatically increasing the yield of rice. The rice plant is able to destroy, and so rapidly detoxify, the herbicide.

The herbicides are usually sprayed on the area to be treated from pressurized containers carried by one man, or from a tank truck, or from aircraft if the area to be treated is large.

2,4-D ([2,4-dichlorophenoxy] acetic acid) and 2,4,5-T ([2,4,5-trichlorophenoxy] acetic acid) have generally low toxicities to animals. 2,4-D has caused irritation of the eyes and gastrointestinal disturbances, but relatively large doses are required for lethality (in cattle, 1 gram per kilogram of body weight). Absorption through the skin is negligible. 2,4-D is used as a weed control primarily on corn and wheat crops, whereas 2,4,5-T is used in pastures for brush control and right-of-way clearing. The herbicide decays rapidly in soil (30 to 60 days), but it is known to last up to 10 months in low-oxygen water.

In terrestrial situations the effects of 2,4-D and 2,4,5-T on animal populations are indirect: by killing off "undesirable" weeds and other plants or by defoliation, the food supply of large numbers of primary consumers (insects, some small mammals, some birds, etc.) may be destroyed, with disastrous effects on these small-animal populations.

In aquatic environments also the effects of these herbicides on fish is indirect. Treatment of water-fouling weeds usually results in rapid bacterial decomposition of the dead plants and concomitant reduction or total elimination of dissolved oxygen. Oxygen-dependent animals may then die of suffocation.

2,4-D and 2,4,5-T are known to be teratogenic (causing birth defects) in experimental animals. A highly toxic contaminant of 2,4,5-T called TCDD (tetrachlorodibenzoparadioxin), an undesirable by-product of the reaction of 2,4,5-trichlorophenol (p. 145) is also known to be a powerful teratogen, and evidence is accumulating to indicate that it, rather than 2,4,5-T itself was responsible for the effects shown in earlier studies. There is some evidence that 2,4-D is carcinogenic, and it is known to produce chromosomal aberrations in plants in concentrations of 10 ppm. Some of these aberrations alter stature and color of the treated plants, and the changes furthermore are inherited by the next generation.

2,4-D and 2,4,5-T are the two most widely used herbicides in the United States. Pressures are now being exerted by environmentalists to reduce or ban 2,4-D and 2,4,5-T for use in the United States.

FUNGICIDES

Fungicides are designed to be toxic to undesirable fungi (molds). Their chemical structures vary widely from inorganic copper, borate, zinc, and sulfur compounds (copper sulfate, Crag 169, borax) to various carbamates (fenthion, nabam, zineb) and phthalimides (captan and folpet). The latter two are used widely in agriculture on more than fifty different crops. They are used in the treatment of soil and seeds to reduce fungus contamination and are sprayed directly on leaves. Captan is also used as a dip to prevent fruit rot due to fungus infection. In one extreme case captan was used against peach scab and increased the yield from ½ ton per acre of poor-cull peaches to 7 tons per acre of good-fancy peaches—an increase of 1300%. Total fungicide production in the United States in 1966 was nearly 167 million pounds. In that year 7.5 million pounds of the two phthalimides, captan and folpet, were used on a variety of crops in the United States.

The phthalimides apparently do not break down immediately in nature—cranberries sprayed with captan showed detectable residues of the fungicide 83

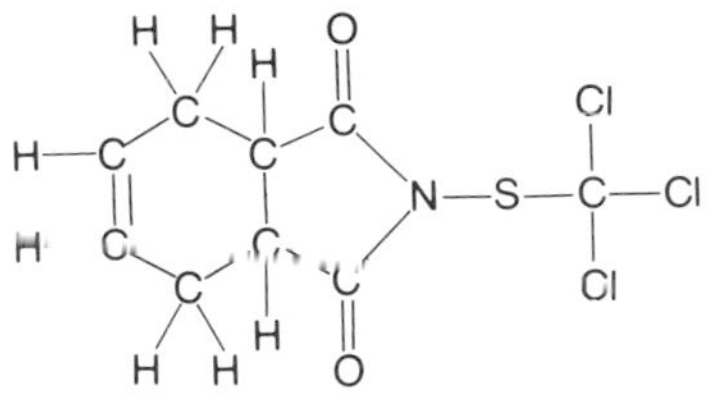

Captan

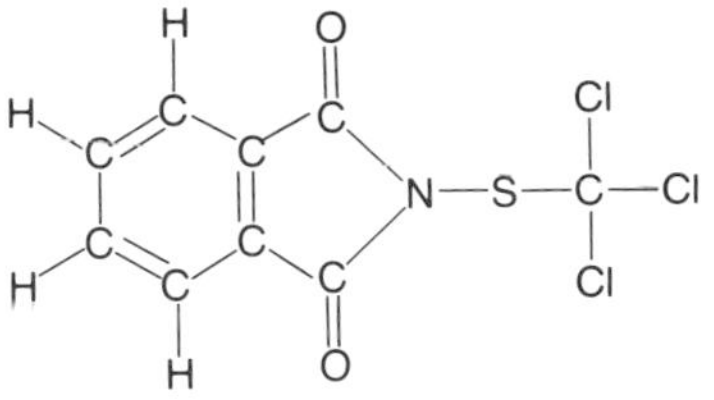

Folpet

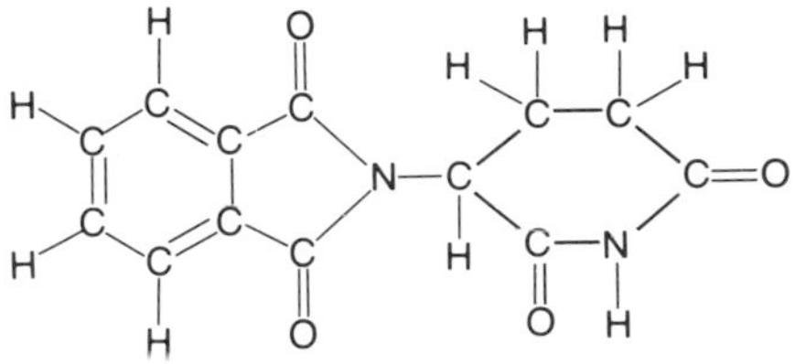

Thalidomide

days after treatment. Prunes dipped in a 0.48% solution of captan showed a residue of 11 ppm even after dehydration of the fruit.

In recent years the phthalimide fungicides in particular have commanded the attention of many health authorities in the United States. Although this class of compounds demonstrates a very low toxicity in mammals when taken orally, they have been shown to cause tumors (i.e., they are carcinogenic), to result in genetic damage (mutagenic), and to cause birth deformities (teratogenic). One of the reasons for the current alarm over captan and folpet is the similarities in chemical structure and action to the compound thalidomide.

During the late 1950's and early 1960's thalidomide was identified as the causative agent in a number of cases of deformed babies born of mothers who had been exposed to the compound. The structural similarities of captan, folpet, and thalidomide are obvious from a glance at their chemical architecture shown above. Thalidomide was formerly used as a sedative until its teratogenic effect was discovered. It was withdrawn from the market immediately.

Teratogenic experiments with captan and folpet on rats and rabbits yield confusing results. In some of the earlier experiments (1968, 1969) they elicited no teratogenic effects. Significant reduction in litter size in rabbits and hamsters (but not in rats) was noted. Later experiments (1970, 1971) showed captan to be teratogenic in New Zealand white rabbits, and captan, folpet, and a related phthalimide difolatan were teratogenic in chick embryos at doses of 3 mg per kilogram of body weight. In a 1970 experiment on golden hamsters, 1000 mg per kilogram of body weight produced high fetal mortality and eight deformed fetuses out of thirty-five born to three mothers. In a 1971 experiment using captan, folpet, and difolatan on pregnant rhesus monkeys at 10 to 75 mg per kilogram of body weight, no malformed offspring were produced.

Although these fungicides have been demonstrated to be teratogenic, the results of such experiments have been inconsistent, showing confusing dose-effect responses. Furthermore, none has been shown to be teratogenic in man, and there is no certainty that they are. Exactly because of their demonstrated teratogenicity, they must not be ruled out in the human case. The site or sites of biochemical action of these teratogens is not known; these compounds therefore invite intensive further research.

OTHER PESTICIDES

In addition to insecticides, herbicides, and fungicides, many other compounds are available that are designed to be lethal to other groups of organisms. These include rodenticides such as Antu (alpha-naphthylthiourea), pindone (2-pivaloyl-1,3-indandione), strychnine (an alkaloid plant extract) and warfarin (3-[alpha-acetonylbenzyl]-4-hydroxycoumarin); molluscicides such as Zectran (a carbamate); nematocides such as Zinophos (a phosphosulfur compound) and Fumazone (a bromine and chlorine compound), and a variety of soil fumigants, wood preservatives, and animal repellents. Most of these products are generally of low toxicity to humans at the concentrations usually employed.

PERSPECTIVE

There is no question that the widespread use of insecticides and herbicides has greatly increased crop yields and so has contributed to elevating the standard of living. The chlorinated hydrocarbons (e.g., DDT) have also contributed significantly to reducing certain insect-borne diseases such as malaria. The chlorinated hydrocarbons have now been used—and used somewhat indiscriminately—for over 25 years, and some troublesome aspects have cropped up. They are highly toxic and very persistent in the environment. There may be no living thing on the ground or in the air (and perhaps in all the earth's waters) that does not contain one or more of the chlorinated hydrocarbons in its tissues. Also, the generally unforeseen circumstance of resistance to these chemicals has occurred. Some strains of malaria mosquitoes and houseflies have developed that are completely refractory to DDT.

It appears that man has subjected the earth and its life to an unacceptable burden of these compounds and, with second thoughts, the use of DDT has been banned in the United States, except for a few special applications. Dieldrin and aldrin may soon follow.

The organophosphates may be considered the "second generation" of synthetic insecticides. These have inherited from the chlorinated hydrocarbons the current responsibility for insect control. Although often more toxic, they are less persistent so do not pose, it is thought, a long-term environmental hazard. It should be realized, however, that any "synthetic" control over living elements in the world may pose a hazard to the natural balances of the environment. These natural, and delicate, balances are the result of millions of years of co-evolution among animals, plants, and the physical factors in ecosystems. These second generation pesticides are short lived and thus localized in their effects. Even so, the eradication or reduction in numbers of one form in a food web could have far-reaching effects on many other levels. If only small forms are involved, their reproductive potential could restore the balance quickly. For larger forms the balance in numbers between different species could be shifted permanently. Man, at the top of several food webs, will inevitably be affected by such shifts at lower levels.

A newer, "third" generation of insect control is now being investigated. This is the use of insect hormones in reducing the populations of harmful insects. Such hormones have several distinct advantages over the highly toxic synthetic chemicals. First, they are required at certain times during insect development, from egg to larva to adult, and must be absent at other times. Metamorphosis from larva to adult can be accomplished only in the absence of the juvenile hormone, for example. It must also be absent from the fertilized egg, or abnormal (and lethal) development results. Second, these hormones are highly insect-specific; they appear to have no effect on other animals or on plants. Third, the controlling effects of the juvenile hormone, in particular, are impressive: as little as 4 grams (less than 1/5 ounce) of a synthetic derivative of the hormone is believed to be potent enough to eradicate all insects from a 1-acre area (a square plot of land 208.71 feet on each side). The pure insect hormone is many times more active. Fourth, insects cannot develop resistance to their own hormones—to do so would be self-destruction. The hormones now known are not species-specific: they affect all insects. There is known to be at least one hormone-derivative, however, that may be specifically lethal to one insect *family,* and this is getting close to the ideal. It appears that man's battle with the harmful insects may be won by using such natural biochemical missiles.

Many authors have pointed out the necessity of an integrated control program, where both biological and chemical controls are meshed to afford maximum protection to the environment. This was not done in the heyday of the chlorinated hydrocarbons. It is not being done today in the ever-widening use of the organophosphates and herbicides. Dr. Charles Wurster of the State University College of New York at Stony Brook wrote the following:

> Modern agriculture must adopt effective, economical and ecologically sound, integrated insect pest management systems to avoid the numerous shortcomings, hazards, and high costs of complete reliance on insecticides. An increasingly hungry and polluted world can ill afford to continue on its present course; if it does, the adaptable insects will be the ultimate winners.

SUGGESTED READING

Finkelstein, H.: Preliminary air pollution survey of pesticides: a literature review, prepared for U. S. Department of Health, Education, and Welfare, Raleigh, N. C., Oct., 1969, National Air Pollution Control Administration.

Shea, K. P.: Captan and folpet, Environment **14**(1):22-32, 1972.

Williams, C. M.: Third generation pesticides, Scientific American **217**(1):13-17, 1967.

Wurster, C. F.: Chlorinated hydrocarbon insecticides and the world ecosystem, Biological Conservation **1**(2):123-129, 1969.

Wurster, C. F.: Effects of insecticides, reprinted in Congressional Record, no. E8333 to E8337, July 28, 1971.

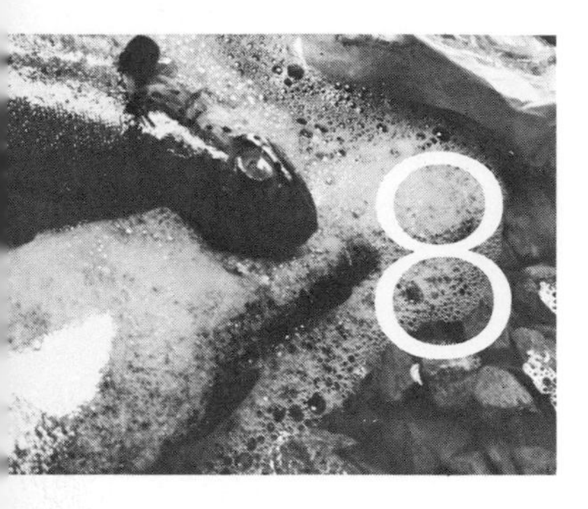

8 MISCELLANEOUS POLLUTANTS

It was inevitable that several important groups of pollutants would not fit well into a classification such as the one that was adopted for this book. A final miscellaneous chapter provides a convenient place to put them. It contains strange bedfellows. Who would expect to find a discussion of food additives followed by one on municipal sewage; on the other hand, what could be more logical? Strictly speaking, the inclusion of a section on radiation requires a slight stretching of our title *Chemical Villains.* This is, however, one of the most important classes of pollutants that man will have to deal with in the near future, and so the subject deserves a place in the book. A short discussion of "the newest pollutant," heat, has been included in the radiation section because it has taken on real importance in connection with the installation of more and more nuclear power plants. One significant air pollutant, noise, has been omitted entirely because to include it would have required stretching the title to the point of letting in other topics that are more properly treated in other books in this series.

FOOD ADDITIVES

Description and uses. The topic of food additives has engendered much controversy in recent years. The authors of books like *The Chemical Feast, The Poisons in Your Food, Food Pollution, The Great American Food Hoax,* and *Consumer Beware* have convinced many of their readers that anything added to food to improve appearance, texture, or flavor is automatically bad. They campaign vigorously for "natural" foods—those without any artificial additives. On the other side of the issue are the food products manufacturers, who say that these additives are necessary to make their products economically viable. Without them the food would either spoil on the shelf or be too unattractive to the consumer and hence unsalable. The poor consumer is caught in the middle with the health of his body and the health of his pocketbook at stake.

Defining food additives is no easy task. The category includes everything from iodine, added to table salt to prevent simple goiter, to diethylstilbestrol (DES), a hormone fed to cattle to increase their rate of growth. In between are more than 2000 compounds that are added to one or another kind of food to color it, preserve it, flavor it, alter its texture, or perform some other task (Fig. 8-1). The rapid expansion of the so-called convenience food industry has been made possible only with the aid of such additives. If by some form of legislation all additives were henceforth prohibited, the majority of urban Americans would suffer an immediate famine, from which many would never recover. Chemical preservatives are particularly necessary to allow time for the food to reach the consumer from the processing plant before it spoils. Seeing "benzoic acid" or "sodium benzoate" on the label of a can of fruit juice causes little fear in the heart of the average consumer, and

Fig. 8-1. Selection of ingredient panels from boxes on the pantry shelf. How many of the ingredients listed on each label are really food substances and how many are additives?

the presence of this preservative makes sale of the product possible. Finding such things as butylated hydroxyanisole (BHA) and butylated hydroxytoluene (BHT) listed as ingredients in potato chips and other oil-containing foods might cause somewhat more concern. Knowing that these materials serve an antioxidant or rancidity-preventing function is usually enough to convince the shopper to buy the potato chips in spite of the imposing-sounding chemicals that they contain. It might be possible to overcome the esthetic problems created by the lack of coloring, flavoring, and texturizing additives, but even here there is cause for concern. One need only observe the number of protein-starved people in South America who refuse to eat Incaparina (a high-protein food supplement supplied either free or at very low cost) because it is foreign to their normal diet. Great difficulties have also been encountered in gaining acceptance for the new strains of "super rice." This resistance seems to revolve around the fact that the new rice grains do not adhere to each other as the old strains did, making the new rice harder to eat with the fingers. One can envision great difficulty for the manufacturer who tries to market a new brand of oleomargarine that is white instead of yellow.

A section titled "effects on organisms" covering the food additives would have to be much longer than this entire book, or even the entire series. Rather than attempting such a treatment, the difficulties surrounding the use of a single additive will be examined. This will serve as an example of the problems involved in the regulation and use of food additives, both for the producer and the regulatory agency involved. The effects of another additive, nitrite, are treated in some detail in Chapter 5.

The case of diethylstilbestrol (DES). Diethylstilbestrol is a synthetic female sex hormone that has been used by meat and poultry producers as an aid in speeding the fattening of their animals. It was used in chickens in the early 1950's to produce caponettes, a tender-meated bird similar to a capon. In the case of the caponettes DES in the form of a pellet implanted in the neck was used in lieu of castration, the technique required to produce capons. In 1954 DES was approved for use in cattle feed with the proviso that it be withheld for the last 48 hours prior to slaughter to allow the material to be cleared from the animal's system.

The objection to the use of DES in the production of meat for human consumption is that it is known to be carcinogenic. Carcinogenesis has been demonstrated in test animals, and a particular type of vaginal cancer in humans has been linked to administration of DES to the mothers of the women involved during their pregnancies. In these cases the drug had been administered to prevent miscarriages. Known carcinogens are banned from human food by the Delaney clause in the Food and Additive Amendment of 1958. It states that no substance that "is found after tests which are appropriate . . . to induce cancer in man or animals" may be added to, or be present, in food offered for sale. Most of the additives in use at the time the Food and Additive Amendment was enacted were placed on a list of substances "Generally Recognized As Safe," thereby avoiding the havoc that would have ensued had all such materials been banned until adequately tested. Many substances originally on this GRAS list (e.g., cyclamates) have been found to be carcinogens and so are no longer allowed in food. Current interpretations of this regulation are that no amount, however small, of any known carcinogen is allowed. This, then, is the crux of the whole additive question with regard to carcinogens. As detection techniques improve, carcinogenic substances can be found in more and more kinds of foods. The quantities involved are becoming vanishingly small, but the instruments can detect them. The question now becomes the following: how much of such substances is required to cause cancer and how much less than that amount can be tolerated in food with minimal risk to the people eating it? This question is a difficult one, and there is presently no concensus on the answer. Many cancer researchers believe that there is no "no effect level" of carcinogens and that they should not be tolerated in any quantity whatsoever. Many other equally reputable scientists take the opposing view that such compounds can be ingested in small quantities without undue risk. So far the more conservative view of banning the compounds completely seems to be preferred. How does all this controversy apply to DES?

When DES was first authorized for use in cattle feed, analytical techniques were not sensitive enough to detect it in meat from animals that had been treated with it. In 1959 the detection of DES improved to the point that it could be found in poultry that had been exposed to it so its use in chickens was prohibited. In 1962 the Delaney clause was amended to, in effect, eliminate DES from regulation. This clause states that farmers can feed carcinogenic substances to meat animals as long as no residues are left in the meat when the chemical is used according to package directions that are "reasonably certain to be followed in practice." This put the onus for finding DES in meat not on the manufacturer of the chemical but on the farmer. Even at this juncture the U. S. Department of Agriculture was not looking for DES in meat. When they did start to check for it regularly, they found it in a few of the samples (1% to 2%). Their technique was sufficiently sensitive to identify DES at levels down to 10 ppb. It was known at that time that mice fed DES at 6.5 ppb developed tumors so that, with this technique, meat testing free of the chemical could still contain enough to cause trouble. The number of samples checked in the early years of such testing (Table 8-1) were hardly adequate to reflect the incidence of DES in the 30 million cattle butchered each year. Table 8-1 shows a decrease in number of samples checked for DES until 1971, when 6000 were examined. These were declared to be free of DES when the records showed that ten of the animals had DES in them at levels as high as 37 ppb. On investigation it was found that a lower official of the U. S. Department of Agriculture had ordered these results suppressed until confirmed by a second method of analysis. Since no second method was available, the positive results were not reported. The battle to ban DES went through several more rounds before action was taken in August, 1972, to restrict its use in cattle-feeding operations. As of January 1, 1973, DES was no longer allowed in cattle feed. It will still be possible to use DES as a pellet implanted in the cow's ear. This use is continuing to be allowed because "USDA has never detected a residue when implants were used as the sole source of DES." The U. S. Department of Agriculture does not regularly sample for DES in animals that have been treated by implant so that it is not surprising that they have not found it in such animals. But even this statement is less than accurate because a DES residue of 60 ppb was detected in a sample of liver taken from a cow that had been exposed to DES only by the implant route.

TABLE 8-1. Number of samples tested for DES by the U. S. Department of Agriculture

YEAR	NUMBER OF SAMPLES
1966	1023
1967	495
1968	545
1969	505
1970	192
1971	6000

The quantities of DES so far detected in beef have admittedly been small. The government agencies responsible for regulating its use, however, have heard repeated testimony from researchers at the National Cancer Institute and elsewhere to the effect that the DES residue levels found in beef should be reduced or eliminated.

Perspective. DES has been cited here as an example of the controversy presently raging over the use of food additives. The same kinds of difficulties may be encountered with many more of the materials now tolerated in foods—one need only recall the

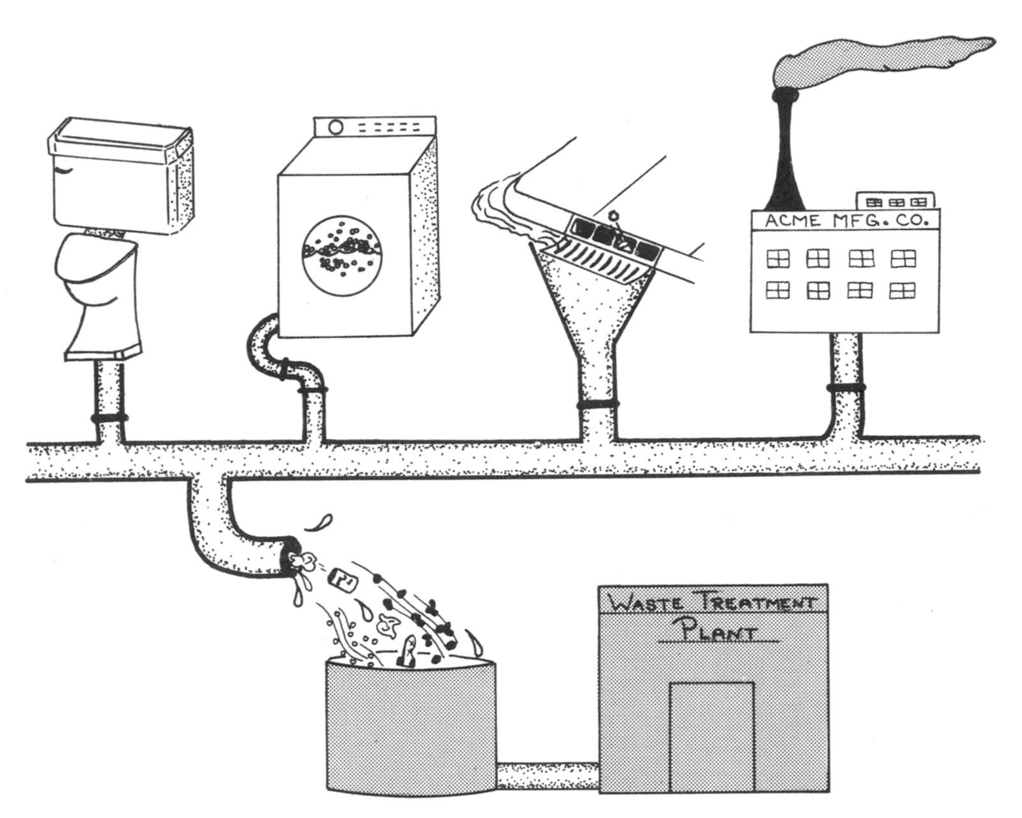

Fig. 8-2. Municipal sewage problem. The waste stream entering the typical sewage treatment facility is so complex that adequate treatment is a near impossibility.

cyclamate battle of the late 1960's. No one seriously favors banning all additives from all foods. On the other hand, such materials should be examined closely for signs of detrimental effects. Before introduction a new additive must be subjected to an extensive (and expensive) battery of animal toxicity studies. Current estimates of the cost of testing a new additive run up to $100,000. Such testing is also being conducted on many of the additives that were in use before testing was required, and some of these (e.g., cyclamates) have been found to fail the tests and are no longer allowed in foods.

In summary, man has been adding extraneous materials to his food for a variety of purposes for thousands of years, and he seems likely to continue to do so in the foreseeable future. It seems prudent to examine the chemicals involved in such practices and to discard those that may threaten his health. On the other hand, there does not seem to be anything wrong with the practice as a whole, and in fact, it would be foolish to advocate a total ban on food additives. Most modern food technology requires the use of certain food additives, and banning them would place an unacceptable strain on the entire food segment of the economy—not to mention making the average diet much less palatable and interesting.

Fig. 8-3. Typical secondary sewage treatment plant. (Courtesy Indianapolis Department of Public Works, Indianapolis.)

MUNICIPAL SEWAGE

Description and occurrence. The invention of the flush toilet, reputedly by the Englishman John Crapper in 1870, constituted the largest single contribution to present water quality problems. The vast quantities of water required to operate them, together with the maze of pipes constructed to carry away the effluent (Fig. 8-2), are a monument to the Victorian notion that bathroom events should be kept out of sight and mind. This attitude has continued to the present day, and it has been broadened to include all manner of wastes, thereby exacerbating environmental problems.

Municipal sewage, like crude petroleum, is a complex and variable mixture of organic compounds, but in this case, a man-made one. In addition to human excreta, municipal sewage contains soaps and detergents, paper, and the effluent from many kinds of industries. It may also contain dead animals, old shoes, tin cans, and anything else that an enterprising child can push down a storm sewer. When it was first invented, the purpose of a municipal sewage system was to collect the wastes from homes and businesses and "transport them out of town." At first this meant dumping the collected material into the nearest available stream or lake. As population density increased and water quality deteriorated, attempts at treating the effluent before dumping it began.

Modern sewage treatment is a two-step process, involving screening and sedimentation to remove suspended solids (primary treatment), followed by some form of biological treatment to remove the dissolved organic material (secondary treatment) (Fig. 8-3). The biological stage uses natural decomposer organisms (bacteria, worms, protozoa, etc.) to break down the complex organic material contained in the sewage into simpler inorganic molecules. The effluent from such treatment is discharged into a receiving stream or lake, where it is hoped that the natural decomposer organisms will break down the last 10% to 20% of the waste without undue damage to the environment. As population growth puts greater and greater demands on the available receiving waters, even secondary treatment is inadequate to protect water quality. Efforts must be made to remove almost all (95% to 99%) of the wastes present in sewage. To do this requires a third stage of treatment (tertiary treatment) that is considerably more expensive than the normal secondary treatment (Fig. 8-4). Several possible methods (reverse osmosis, activated carbon filtration, chemical precipitation, etc.) are being investigated, and different ones may be found desirable for different communities.

All of these methods for tertiary treatment of sewage have a high cost factor as well as a common end result, that is, the production of an effluent which, with chlorination, is of drinking-water quality. Many treatment plant operators question the necessity for this level of treatment in their particular situations because the quality of the local streams is already so low. They argue that to put such high-quality effluent back into the streams will have the effect of *improving* water quality, and they see little justification for charging the taxpayers for this service. The logical alternative is to deliver the effluent from the sewage treatment plant directly to the water treatment plant for reuse. This approach might require a strong public relations effort initially, but in terms of dollars it makes sense. A similar approach is favored by the legislators, who have proposed requirements that both municipalities and water-using industries be required to locate their intakes downstream from their discharge pipes. If this kind of regulation were put into effect all over the United States, there might be an end to the day when, so the story goes, the people in Louisville can detect, 3 days afterward, that the supermarkets in Cincinnati (90 miles up the Ohio River) have had a sale on broccoli!

Even if you do not go swimming or fishing, this process of natural rejuvenation of polluted water is of prime importance to *you*. The cleanup process is accomplished by tiny microorganisms (bacteria, protozoa, fungi, algae, etc.) that break down pollutants into nontoxic compounds such as carbon dioxide and water. These microorganisms are killed by high concentrations of some toxic chemicals, and without them, the pollutants added to a stream by one town would appear unchanged in the drinking water of the next town downstream. The total load of pollutants in the stream (and therefore in municipal water supplies) would increase all along the way. The water in the Ohio River has passed through the water systems (and sewers) of at least three, and probably seven or eight, cities or towns by the time it reaches the Mississippi River at Cairo, Ill. How would you like to live in Cairo, Ill., if all the microorganisms in the Ohio River were wiped out?

Effects on organisms. The water leaving a typical secondary sewage treatment plant contains some dissolved and suspended organic matter and large quantities of nitrogen and phosphorus. When this water enters a stream or lake, decomposer organisms immediately attack the organic matter and begin to

Fig. 8-4. Tertiary sewage treatment plant operating in St. Louis County, Missouri. This particular self-contained facility performs complete sewage treatment (primary, secondary, tertiary) continuously for a population of 2500 persons. The discharge water is suitable for recycling as drinking water. (Courtesy Chem-Pure, Inc., St. Louis.)

break it down into inorganic compounds. This process requires oxygen and thereby depletes the dissolved oxygen supply of the receiving water. If too much of the available oxygen is used in decomposition of the sewage, fish and other organisms may die for lack of it. The nutrients supplied to the receiving water by the treatment plant effluent are also instrumental in promoting eutrophication (Chapter 5). In either case the effects of discharging this treated waste into a stream or lake are proportional to the relative volumes involved. If the municipality is small and the receiving stream large, little damage will result from such discharge. If, on the other hand, the situation resembles that in Indianapolis, difficulties are almost sure to arise. In this case the receiving stream is the White River, and the volume of effluent leaving the city's two treatment plants frequently equals or exceeds that of the stream during the summer low-flow periods. In this situation a treatment plant that is 90% efficient in removing wastes will not be adequate to protect the water quality of the receiving stream, whereas in cases where the receiving stream is very large, removing 75% of the wastes might result in a cleaner river below the treatment plant.

In addition to human wastes and their breakdown products the effluent from treatment plants usually contains a variety of industrial chemicals (phenols, dyes, pesticides, etc.) that are not broken down during secondary treatment but pass through it unaffected. They may then cause problems for sensitive organisms downstream. Some of these chemicals arrive at the treatment plant in sufficient concentration to temporarily destroy the decomposer organisms, thereby inhibiting the second stage of treatment. When this happens, the waste stream passes on through the plant with only primary treatment until the decomposer populations can be reestablished. Incidents of this type are occurring with increasing frequency in many municipalities.

Perspective. The municipal sewage problem is one of the largest facing the United States today. It is receiving increasing attention at all levels of government, but the problem is still growing faster than the efforts to control it. During the period from 1957 to 1970, some $6.4 billion were spent on construction of new treatment plants, with a larger amount allocated to operating costs. President Nixon requested $2 billion for construction of new treatment facilities in the fiscal year 1972, but the Congress appropriated only $1.65 billion. The Council on Environmental Quality estimates that the United States state and local governments will need to spend between $36 billion and $75 billion for construction of treatment plants and combined sewers between 1971 and 1980 if water quality is to be brought in line with present standards. The 1972 report of the Council on Environmental Quality indicates that whereas overall air quality improved during 1971, water quality is still deteriorating. Only if people are willing to devote significant sums of money and serious research effort to this area is the downward trend in water quality likely to be reversed.

RADIOACTIVE MATERIALS

Description, natural occurrence, and production sources. Since the end of World War II the term *radioactivity* has become virtually a household word. Natural radiation has been known at least since 1895 when x rays were discovered. Very soon thereafter serious biological effects of radiation were encountered as radioactive materials were purified and concentrated.

Today with the ever-increasing use of x rays in medicine and radioactive materials, particularly in the generation of electric power, radiation has become a potential health hazard. (Before the moratorium on nuclear weapons testing in the atmosphere this hazard was of global proportions.) So far adherence to strict procedures in the handling of radioactive material has permitted a nearly unblemished safety record.

Description. Radiation is a property of a small proportion of certain atoms that make up all matter. These atoms (the smallest part of an element) are composed of a central nucleus having a positive electrical charge surrounded by a cloud of particles (electrons), each with a negative charge. The nucleus itself is the source of radioactivity. It is composed of two groups of major particles: protons, each of which has a positive charge, and neutrons, which have no charge.

Ordinarily the number of protons in the nucleus equals the number of electrons outside the nucleus, which makes the atom electrically neutral. Most atoms of an element have a specific number of protons and neutrons making up the nucleus. Some proportion of the atoms of certain elements (15 of the 103 known elements) are found in nature with a different number of neutrons in the nucleus, and these 15 elements show radioactivity. They include carbon, hydrogen, potassium, lead, radium, and uranium. Some 65 more elements can be made radioactive. They include calcium, cobalt, copper, iodine, nickel, phosphorus, and sulfur.

The radioactivity is due to a "decay" (breaking

TABLE 8-2. Half-life and type of irradiation for several radioisotopes

RADIOISOTOPE	HALF-LIFE	TYPE OF RADIATION
Silver 110	24 seconds	Beta, gamma
Copper 64	12.8 hours	Beta, positron, gamma
Iodine 131	8.05 days	Beta, gamma
Phosphorus 32	14.2 days	Beta
Hydrogen 3	12.3 years	Beta
Lead 210	19.4 years	Beta, gamma
Radium 226	1620 years	Alpha, gamma
Carbon 14	5600 years	Beta
Uranium 238	4.5 billion years	Alpha, gamma

down) of the nucleus of these atoms. An element composed of atoms having this unusual number of neutrons is called an isotope; radioactive isotopes (not all isotopes are radioactive) are radioisotopes. The radiation may be due to emanation of electrically charged particles or noncharged electromagnetic energy. These may be positively charged (*alpha* particles or positively charged electrons called *positrons*) or negatively charged (*beta* particles, which are electrons). The electromagnetic radiations are called gamma rays. It is possible to identify the radioisotopes not only by their type of radiation but also by the *rate* at which they decay. The term *half-life* specifies the time required for a particular radioisotope to release one half of all its radioactivity; this value varies from seconds to many years, and it is very specific for each isotope. Table 8-2 shows the half-life and type of radiation for several radioisotopes. The number of each element in Table 8-2 is the atomic weight (mass) of the isotope, which is the combined mass of the neutrons and protons in the nucleus. Each neutron and each proton is assigned a mass of one.

Like any potential environmental pollutant radioactivity must be measured if it is to be controlled. The basic unit is the curie (Ci), which is the rate of decay of 1 gram of radium. The unit in which radiation exposure is measured is the roentgen (R), which corresponds to an energy absorption of about 88 ergs per gram. Two additional units of radiation are useful. The rad (radiation absorbed dose) is defined as the deposition of 100 ergs of radiation energy. The absorbed dose in rads varies in different materials for the same exposure in roentgens. An exposure of 1 roentgen of gamma rays produces an absorbed dose of about 0.97 rad in muscle. The rem (roentgen equivalent mammal) is a unit that defines the effect of a certain type of radiation on living tissue compared to the effect produced by ordinary x rays. Since different types of radiation produce different biological effects (for a given amount of energy delivered to the tissues), another unit must be employed to relate rads to rems. This is the relative biological effectiveness factor (RBE). The RBE for x rays or gamma rays is 1, for beta particles it is 1, and for alpha particles it is 10 to 20. The relationship between rads and rems is as follows:

$$\text{Dose rems} = \text{RBE} \times \text{Absorbed dose, rads}$$

Several types of precision instruments are available for the measurement of radiation.

Natural sources. Natural or "background" radiation constitutes about 68% of all radiation to which humans are exposed. This radiation comes from gases emanating from radioactive minerals in the earth's crust and from the action of cosmic rays on atmospheric gases. The most important of the radioactive gases from the earth's crust is the noble gas radon 222 and its decay products. Radon 222 is an alpha-emitter with a half-life of 3.82 days. The atoms of radon and its products attach to dust particles and so can be transported in the atmosphere. The interactions of cosmic rays on atmospheric gases produce such isotopes as carbon 14, hydrogen 3, and beryllium 7. Gamma rays, electrons, and certain other radiations are produced by these isotopes. Uranium and radium deposits also contribute to the total surface radioactivity.

Another source of natural radioactivity is from the burning of fossil fuels (coal, gas, oil). Coal contains about 1 ppm of uranium 238 and about 2 ppm of thorium 232. The major radioactive pollutants in the ash from burning coal appear to be radium 226 and radium 228, which originate from the uranium and thorium. Oil also contains radioactive elements, and radium 226 and radium 228 can be detected in its combustion products. Radon 222, a product of radium 226 decay, is found in natural gas but in very low quantities (measured in trillionths of a curie per liter of gas).

Production sources. Radioactivity in the atmosphere is also due to activities of man in this nuclear age. The radioactive pollutants are the same whether they originate from mining and processing of uranium and thorium ores, or from the operation of nuclear power reactors, or from atomic and hydrogen bomb detonations. Many radioisotope pollutants produced are in the form of gases. Some of these are listed in Table 8-3 with their sources and half-lives.

TABLE 8-3. Half-life of some radioactive gases and their production sources

RADIOISOTOPE	HALF-LIFE	SOURCE
Radon 222	3.8 days	Mines, refineries
Iodine 131	8 days	Reactors, bombs, fuel processing
Sulfur 35 (as SO_2)	87 days	Laboratories
Krypton 85	10 years	Reactors, fuel processing
Carbon 14 (as CO_2)	5600 years	Bombs, laboratories

Radioactive wastes may be retained by the fuel elements of a reactor, or the coolant used may become radioactive during reactor operation. Such radioactive wastes represent potential pollutants if released to the environment during, for example, the reprocessing of spent fuel. Activation and subsequent release of the coolant liquid could add additional radioactivity to the atmosphere or waters.

Nuclear weapons testing before the test ban of October, 1963, added measurable amounts of radioactivity to the atmosphere (Fig. 8-5). The major fission products of a nuclear explosion are strontium 90, cesium 137, and iodine 131. The total explosive yield of all nuclear bombs through the end of 1962 is estimated to be 511 megatons (511 million tons), TNT equivalent. These detonations have added from 10% to 15% to the natural background radiation. Medical applications of x rays and radioisotopes currently contribute an average of about 30% of the total human radiation exposure. The remaining 2% of the total radiation is due to fallout from weapons testing, high-altitude jet travel, color television sets, and luminous dial watches, with radiation from nuclear power plants near the bottom of the list.

Occurrence in the environment. As noted earlier, the source of environmental contamination by radioisotopes may be from their use in industry and research (e.g., as "tracers" in biological research), from medical applications (internal irradiation to control tumorous growths), from the nuclear reactor industry, and from nuclear weapons testing. The wastes resulting are generally the unused remains of the original isotope employed. Depending on the half-life of the isotope, the wastes may retain significant levels of activity. At the present time the total quantities used in industrial and medical applica-

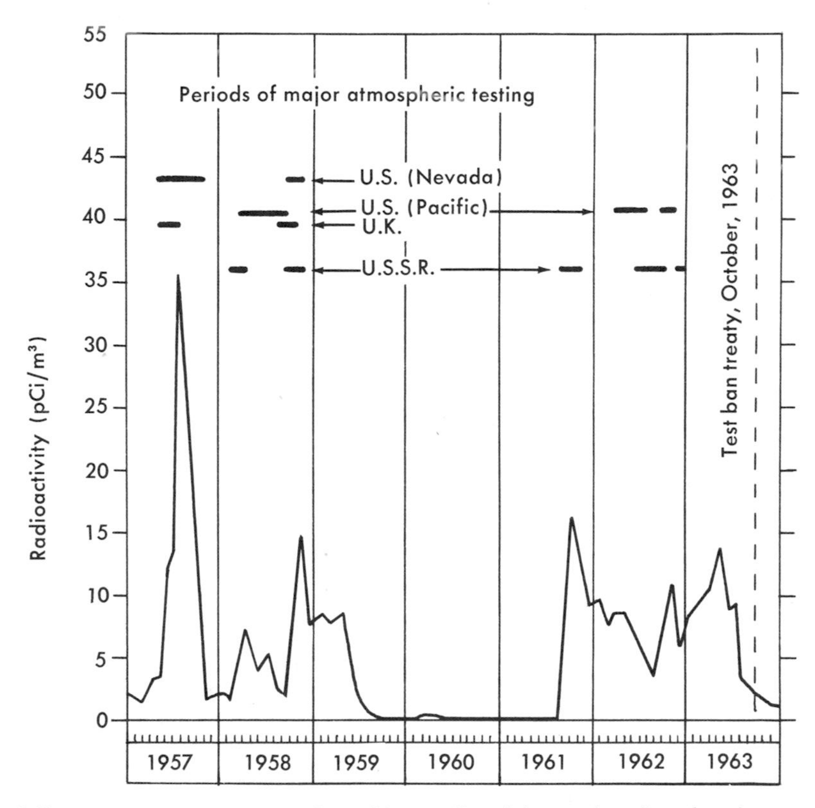

Fig. 8-5. Monthly mean concentrations of beta radioactivity as a function of nuclear weapons testing. Fallout from nuclear testing has been the only significant man-made addition to environmental radiation loads. (From Miner, S.: Preliminary survey of radioactive materials, pub. no. APT D 69-46, Department of Health, Education and Welfare, National Air Pollution Control Administration, Raleigh, N. C., 1969.)

tions are small and have not yet presented an environmental hazard. The use of gamma emitters does not result in radioactive waste accumulation.

Atmospheric levels of radiation are measured at several stations across the United States by the Radiation Alert Network (RAN). This agency is concerned primarily with monitoring beta radiation resulting from weapons test fallout. Fig. 8-5 shows the monthly mean concentrations of beta radioactivity as a function of nuclear weapons testing (atomic bombs and hydrogen bombs). The radioactivity is expressed as picocuries (pCi) per cubic meter (m^3) of air sampled. A picocurie is a trillionth of a curie. Atmospheric radioactivity from bomb explosions has been low since the test ban treaty of 1963. Activity increased only during the French and Chinese tests.

Table 8-4 lists maximum gross beta radioactivity (in pCi/m^3) for selected states of the United States for selected years (1958, 1960, 1963, 1965). The values match quite closely the trends shown by the graph in Fig. 8-5, as might be expected.

TABLE 8-1. Maximum beta radioactivity in pCi/m^3 for selected states and selected years*

	1958	1960	1963	1965
California	126.0	0.9	31.2	1.4
Washington	18.0	0.3	14.3	0.8
Montana	508.0	0.4	35.2	1.4
Oklahoma	17.0	0.4	24.5	1.3
Michigan	16.6	0.4	17.2	1.5
Alabama	20.3	0.4	16.6	1.1
Maine	29.0	0.4	15.8	1.3
New York	20.6	0.4	25.2	1.2
Florida	39.0	1.2	33.0	—
South Carolina	21.0	0.5	20.0	1.3

*Maximum permissible concentrations vary with the isotope. For strontium 90 it is 30 picocuries per cubic meter (pCi/m^3) for soluble forms, and for iodine 131 it is 100 pCi/m^3 for soluble forms.

Mode of entry and accumulation in the body. Exposure to radiation may be from a distant source that occurs only by gamma radiation or from radioactive particulate matter that may be deposited on the body surface, inhaled, or ingested in food and water. Contamination via the respiratory system (inhalation) is the most direct avenue of entry to the body. To enter the body via the digestive system food and water must first be contaminated. The average radiation dose received per person in the United States is estimated to be between 75 and 175 millirads (mrad) per year from natural sources and in excess of 1200 millirems (mrem) per year from medical sources. If the radiation is beta particles, rads and rems are equivalent. Nuclear industry workers receive from 50 to 100 mrads per year. Fallout from nuclear weapons testing contributes about 3 mrem per year, and all remaining sources contribute less than 1 mrem per year.

Inhaled radioisotopes may remain in the lungs, irradiating that tissue, or may penetrate the alveolar walls and enter the bloodstream. Once in the blood, such radioactive atoms may be transported throughout the body. Different isotopes may be concentrated preferentially in different tissues. The isotopes of strontium are deposited primarily in the bones; carbon and hydrogen isotopes are distributed throughout all body tissues; radioisotopes of iodine accumulate in the thyroid gland; cesium isotopes concentrate in soft tissues, including the gonads. The biological damage due to irradiation by an isotope depends not only on the dose level but also on the half-life of the isotope. For example, comparable doses of iodine 131 with a half-life of 8 days would cause less biological damage overall than would strontium 90 with a half-life of 28 years.

The body makes no chemical distinction between the isotopes of the same element. All are subject to the same metabolic reactions. Radioisotopes may therefore be accumulated for long periods or incorporated into waste products and eliminated in the urine or feces.

Cesium 137 produced by atmospheric testing of atomic weapons is a case in point (Fig. 8-6). Cesium is similar chemically to potassium and is absorbed along with it. Plants seem unable to distinguish between these elements and absorb them in proportion to their availability. Animals eating plants containing cesium 137 also absorb it along with the nonradioactive potassium, but their digestive tracts do absorb more of the potassium than cesium when both are present in the food in equal amounts. Once in the

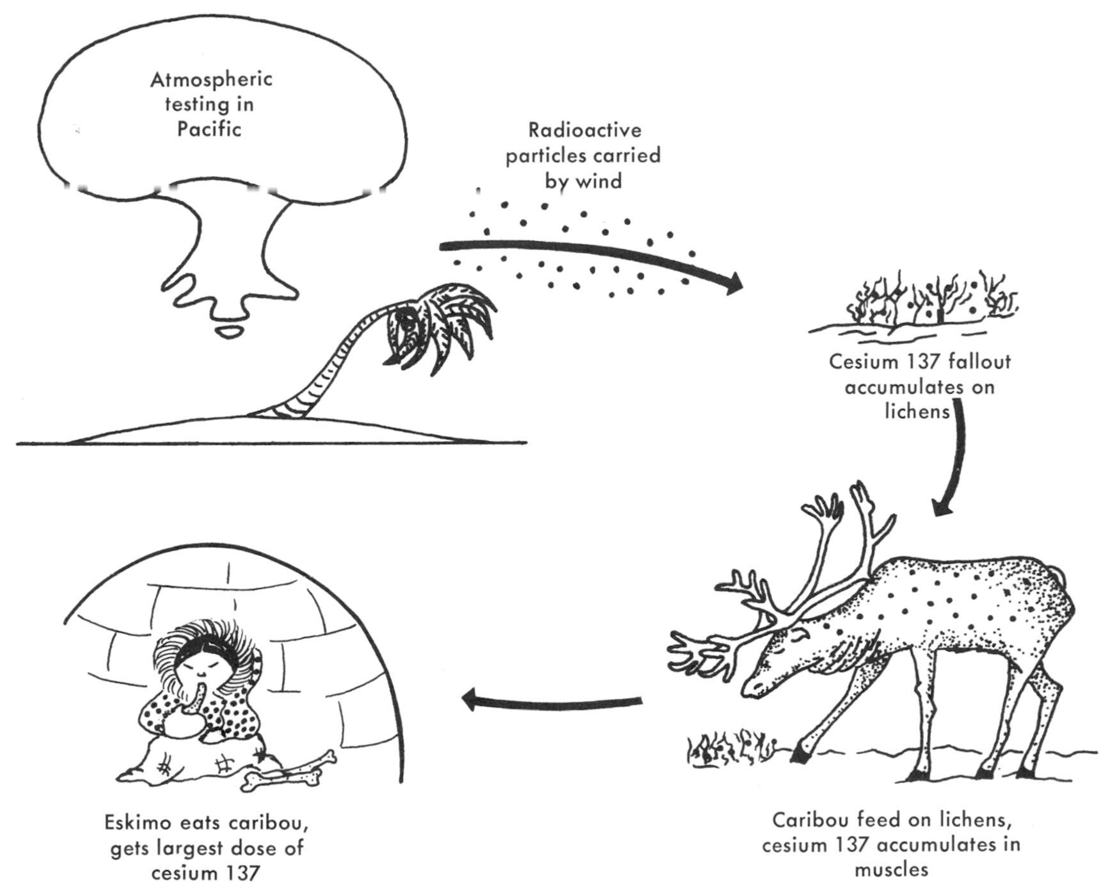

Fig. 8-6. Food chain concentration of cesium 137 in the arctic ecosystem. The black dots indicate increasing concentration at each step in the food chain.

animal, cesium goes to the muscles along with potassium. When radiation-contaminated vegetation is the sole source of food for a mammalian herbivore like caribou or reindeer, the concentration of radioactive cesium in the tissues (meat) of the animal can be twice that of these elements in the plants eaten as food. If the herbivore is eaten by a carnivore (man or another animal), the concentration of radioactive materials can double again. This is precisely the situation that prevailed in arctic regions after the atmospheric testing of the mid-1950's. The lichens growing on the tundra became contaminated by fallout from the testing, and they served as the sole food source for the Alaskan caribou. The caribou, in turn, were almost the only source of food for the Eskimos in winter, and significant increases in whole-body radiation counts occurred among these people.

Biological effects of radiation exposure. The biological effects of high levels of radiation exposure over short periods are well documented. The series of biological events after acute exposure is termed *radiation sickness.* Doses above 200 rem cause vomiting, which is followed by a quiet period of about 2 weeks. Loss of hair may then occur along with diarrhea, sore throat, and hemorrhages of the skin and internal organs. Exposures of 400 to 500 rems produce about 50% fatalities, and at exposures of 1000 rems there likely would be no survivors. Above 700 rems the quiet or latent period may be as short as 1 day. The symptoms of radiation sickness are mainly due to damage sustained by the blood-forming areas of the bone marrow and lymph system.

Levels of exposure that are not generally lethal have resulted in numerous malfunctions and diseases, including skin, bone, and blood cancer (leukemia). Lung cancers exceed expected values in miners who are exposed to radioactive gases such as radon. Exposure of the lens of the eye to x rays, gamma rays, and beta particles at levels above 550 rads may result in cataracts (lens opacity). Experiments with animals indicate that radiation at sublethal doses results in a shortened life-span. The reasons are not known but are associated with acceleration of aging. No suitable data exist for this effect in man. Interference with proper functioning of the body's immunity system is also known. Some of these effects appear only after some lapse of time following exposure. After the nuclear bombing of Japan in 1945 cases of leukemia appeared in about 1 year but did not reach a peak until 7 years after the exposure.

Radiation not only physically destroys tissue structure so that it cannot perform its function, but it may also damage chromosomes and so cause mutations of genes. If such mutations occur in large numbers of somatic (body) cells, the effect may be seen in that victim's lifetime. If the mutation occurs in the reproductive organs (testes and ovaries), such damaged chromosomes may be inherited by the children of the exposed parent(s), and defects will then appear only in the next generation.

Genetic injury to an entire population will depend on the total "load" of such new mutant genes. Therefore a small but mutation-causing dose to which the entire population is exposed will likely result in more long-term genetic damage than will a high level of radiation delivered to a small or isolated group. It is estimated that a dose between 10 and 100 rads during the fertile years of a human being will double the normal mutation rate. If such a dose were delivered to several generations, a new equilibrium of such mutant genes would result, amounting to twice the number present in the original population. Biologists do not know what contribution is mady by natural background radiation to the present mutation rate.

Perspective. Since radioactive substances harmful to the general population are either airborne or are incorporated in the food chain that supplies our nutrients, abatement involves prevention of excessive amounts of radioisotopes from reaching the atmosphere or surface waters.

Such abatement procedures involve reducing isotope emissions from uranium mining procedures (e.g., underground drainage) and emissions from the burning of fossil fuels. In nuclear reactor operations, emissions are controlled by the use of closed-coolant systems and by using high purity coolants to reduce secondary activation products. Airborne contaminants are reduced by filtration, centrifugation, wet collection, and electrostatic precipitation.

Perhaps the least expensive method of abatement is to collect and store radioactive wastes. This method is useful mainly for reducing activity of the short half-life isotopes before release to the environment. A thousandfold reduction in activity requires storage for ten times the isotope half-life. For xenon 133, a radioactive inert gas with a half-life of 5.3 days, reduction by a thousand times of its activity would require storage for nearly 2 months. Long half-life isotopes that cannot be released to the environment are buried in deep shafts underground, where they are left to undergo their characteristic decay.

To date, only atmospheric bomb detonations have resulted in a significant increase in environmental

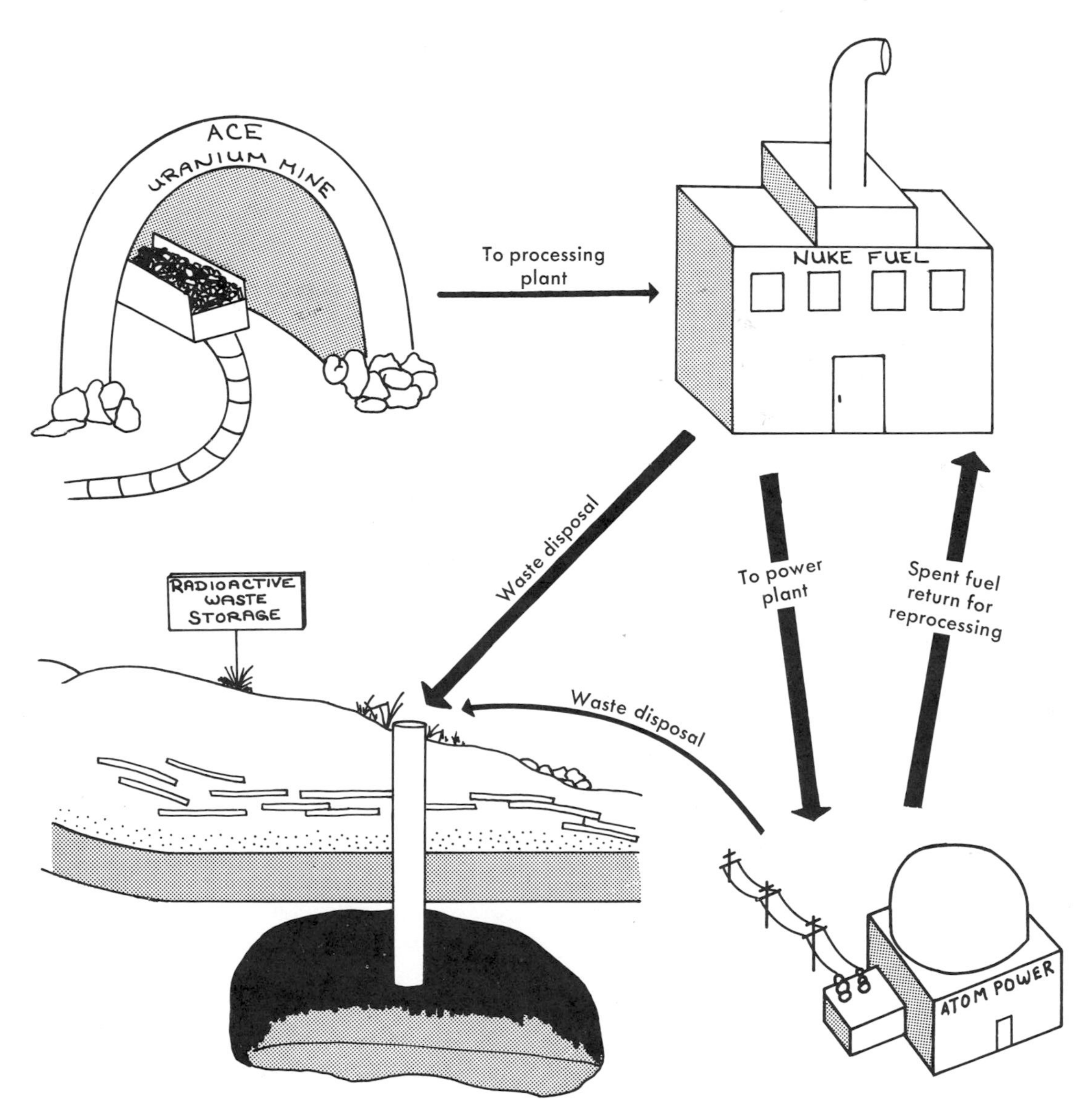

Fig. 8-7. Pathways taken by radioactive materials in nuclear power usage. The fuel reprocessing and waste disposal parts of the cycle present the greatest danger of escape of radioactivity to the environment.

radiation loads, and this increase has only been temporary. Barring a thermonuclear war, such levels probably will not be seen again. Very strict procedures in uranium mining and in the operation of nuclear reactors have so far prevented the release of radioisotopes at levels that can be regarded as hazardous. The increasing emphasis on the use of nuclear reactors as sources of electricity will inevitably increase the likelihood of a serious accident involving the escape of radioactive materials into the environment (Fig. 8-7). The kinds of accidents that can be anticipated for electric power–generating nuclear reactors are not of the atomic bomb variety but, rather, involve the possibility of leakage of radioactive materials into the surrounding air and/or water. This type of accident is potentially possible, both at the reactor site and along routes traversed by vehicles transporting either fuel or radioactive wastes to or from the reactor. The rate of expansion of the nuclear power industry is currently below previous predictions, primarily because serious questions regarding the adequacy of emergency cooling systems are awaiting definite answers. In the meantime other sources of electric power (and systems using hydrogen rather than electricity as a transportable "energy currency") are being explored with renewed vigor.

One aspect of nuclear power plants that has yet to be mentioned is the problem involved with getting rid of the waste heat that they produce. This subject has received considerable attention, but solutions do not come easily. All of the present designs for nuclear power plants involve conversion of heat energy from the nuclear reactions into electrical energy. The Second Law of Thermodynamics (p. 5) states that this conversion cannot be 100% efficient and that some of the energy will be lost as heat. In all present nuclear power plants the losses are approximately 75% of the available energy. This represents 50% to 60% more waste heat per kilowatt of electricity produced than the most efficient fossil fuel–fired plants. This waste heat must be exhausted to the environment. In practice most power plants use water from nearby

Fig. 8-8. Two natural draft cooling towers at the Mitchell Plant of the Ohio Power Company South of Moundsville, West Virginia, on the Ohio River. Each tower serves a generating unit rated at 800 megawatts. (Courtesy American Electric Power Co., Inc., New York, N. Y.)

streams or lakes to carry away the excess heat. With the fossil fuel–fired plants and the early generation of nuclear plants, heating of the surrounding surface waters presented only small localized problems. The future of the nuclear power industry depends, however, on the construction of large plants with correspondingly greater problems in disposing of waste heat. From one of these large power plants the waste heat would be sufficient to raise the temperature of a river carrying 3000 cubic feet of water per second by 10° F. In many streams this amount of heat would be enough to destroy all but the most heat-tolerant species of organisms. It has been estimated that if the nuclear power industry expands at currently projected rates, by the year 2000 one third to one half of the average United States daily stream runoff will be required to cool power plants.

Clearly this level of heating of surface waters is unacceptable, and alternative cooling methods will have to be employed in many situations. Cooling towers have been designed to deal with the waste heat problem by transferring it to the atmosphere (Fig. 8-8). They are expensive to construct and operate, but they have been demonstrated to work successfully and are being installed in many areas. Special artificial cooling lakes are also feasible in some situations. They must be very large (3 miles long by 1 mile wide for a typical 1000-megawatt nuclear power plant), but they are cheaper to build and operate than cooling towers in areas where land is available at reasonable prices. These lakes can also be used for recreational purposes, thereby adding to their esthetic value.

As with all other applications of modern technology the construction of nuclear power plants requires close scrutiny of the possible environmental consequences. Every change in one component of the ecosystem can be expected to have widespread and unpredicted effects on other components. These effects must be considered in determining which action is to be taken in any given situation. To do otherwise is to court environmental disaster.

SUGGESTED READING

Food additives

Alexander, T.: The hysteria about food additives, Fortune **85**(3):63-65, 1972.

Kermode, G. O.: Food additives, Scientific American **226**(3):15-21. 1972.

Wade, N.: DES: a case of regulatory abdication, Science **177**:335-337, 1972.

Radioactive materials

Miner, S.: Preliminary air pollution survey of radioactive substances, pub. no. APTD 69-46, Raleigh, N. C., 1969, U. S. Department of Health, Education and Welfare, National Air Pollution Control Administration.

U. S. Environmental Protection Agency: Questions and answers about nuclear power plants, Washington, D. C., U. S. Government Printing Office.

APPENDIX

UNITS OF MEASUREMENTS USED IN THIS BOOK

MASS

1 gram (g)	=	0.035274 ounce
1 kilogram (kg)	=	1000 grams = 2.2046 pounds
1 milligram (mg)	=	1/1000 g = 0.001 g
1 microgram (μg)	=	1/1,000,000 g = 0.000001 g
1 nanogram (ng)	=	1/1,000,000,000 g = 0.000000001 g
1 pound	=	453,590,000,000 nanograms (ng)

LENGTH

1 meter (m)	=	39.37 inches
1 kilometer (km)	=	1000 m = 0.62137 mile
1 centimeter (cm)	=	1/100 m = 0.01 m
1 millimeter (mm)	=	1/1000 m = 0.001 m
1 micrometer (μm)	=	1/1,000,000 m = 0.000001 m
1 nanometer (nm)	=	1/1,000,000,000 m = 0.000000001 m
1 angstrom (Å)	=	1/10,000,000,000 m = 0.0000000001 m

VOLUME

1 liter (l)	=	1.0567 quart = 1000 cubic centimeters (cm^3 or cc)
1 milliliter (ml)	=	1 cm^3 = 1/1000 liter = 0.001 liter
1 cubic meter (m^3)	=	1000 liter = 1.3097 yd^3

CONCENTRATION

1 part per million (ppm) = 1 μg/g = 1 mg/kg = 1 inch in 15¾ miles
1 part per billion (ppb) = 1/1000 ppm = 1 μg/kg = 1 drop in 13,208 gallons

RADIATION

Curie (Ci) A quantity of radioactive substance that exhibits a disintegration rate of 37 billion disintegrations per second (37×10^9 dps). For radium the quantity is 1 gram.

Roentgen (R) A measure of x rays or gamma rays absorbed by a body, in terms of energy. One roentgen will deliver about 100 ergs per gram to biological solutions. Natural background radiation will deliver about 5 roentgens in 30 years. A dose lethal to about one half of a given human population is 425 roentgens.

Radiation absorbed dose (rad) A unit of absorbed energy equal to 100 ergs per gram. It is about the same value as 1 roentgen but is independent of the substance irradiated.

Relative biological effectiveness (RBE) This value compares the effect of various radiations on biological material to the effect produced by a unit of absorbed dose of x rays. For x rays, gamma rays, and beta radiation RBE = 1; for alpha particles RBE = 10 to 20.

Roentgen equivalent mammal (rem) This is the unit of RBE. The relationship to rad and RBE is as follows: rems = rad X RBE. For x rays, gamma rays, and beta radiation rems = rads. For alpha particles 1 rem = 10 to 20 rads, depending on the energy of the alpha radiation.

INDEX

Boldfaced numbers refer to pages with illustrations; numbers followed by t refer to pages with tables.

D

E

F

U

V

W

Z

9071-15
5-32

628.53
Berry B534c
Chemical villains.